Moisture and Buildir

Woodhead Publishing Series in Civil and Structural Engineering

Moisture and Buildings

Durability Issues, Health Implications and Strategies to Mitigate the Risks

Arianna Brambilla

Alberto Sangiorgio

Woodhead Publishing is an imprint of Elsevier
The Officers' Mess Business Centre, Royston Road, Duxford, CB22 4QH, United Kingdom
50 Hampshire Street, 5th Floor, Cambridge, MA 02139, United States
The Boulevard, Langford Lane, Kidlington, OX5 1GB, United Kingdom

Notices

Knowledge and best practice in this field are constantly changing. As new research and experience broaden our understanding, changes in research methods, professional practices, or medical treatment may become necessary.

Practitioners and researchers must always rely on their own experience and knowledge in evaluating and using any information, methods, compounds, or experiments described herein. In using such information or methods they should be mindful of their own safety and the safety of others, including parties for whom they have a professional responsibility.

To the fullest extent of the law, neither the Publisher nor the authors, contributors, or editors, assume any liability for any injury and/or damage to persons or property as a matter of products liability, negligence or otherwise, or from any use or operation of any methods, products, instructions, or ideas contained in the material herein.

British Library Cataloguing-in-Publication Data
A catalogue record for this book is available from the British Library

Library of Congress Cataloging-in-Publication Data
A catalog record for this book is available from the Library of Congress

ISBN: 978-0-12-821097-0 (print)

ISBN: 978-0-12-821098-7 (online)

For information on all Woodhead Publishing publications
visit our website at https://www.elsevier.com/books-and-journals

Publisher: Matthew Deans
Acquisitions Editor: Glyn Jones
Editorial Project Manager: Hilary Carr
Production Project Manager: Anitha Sivaraj
Cover Designer: Miles Hitchen

Typeset by MPS Limited, Chennai, India

Contents

Foreword

Over the last 50 years, thermal resistance has been the main priority in building performance design and assessment, especially in cold and temperate climates. This trend is reflected in building policy worldwide and design and construction practices, where the increase of thermal resistance of the building envelope is still the main design criteria. However, over the years, a renewed attention for the moisture balance within buildings and their enclosures has emerged, due to the increasing percentage of buildings dealing with mould and other moisture-related issues. The combination of new construction methodologies, with an evident trend towards the use of materials with low vapour permeability (metal cladding, closed-cell foams, etc.) together with stricter airtightness targets are among the main reasons for excessive dampness and mould proliferation nowadays. Meanwhile, new organisational and social approaches that affect the way we occupy buildings more and more continuously have created unprecedented challenges related to the moisture balance in the built environment. Within this context, it does not surprise that moisture is the single main issue affecting indoor air quality and durability issues in buildings and is reported to almost a third of the modern buildings. Although multiple coexisting reasons for the critical moisture levels in buildings and several solutions have been proposed, the topic of excessive moisture levels in buildings remains challenging.

This book provides a critical overview of current research and knowledge about moisture-related aspects. The approach is succinct and has the merit to combine a clear theoretical introduction to the topic together with a pragmatic discussion of modern techniques to deal with it. The achieved goal is a critical assessment of current knowledge of moisture and its implications together with a comprehensive review of the moisture buffer performance of building materials.

In its eight chapters, the book holistically raises the attention for the multifaceted implications of hygrothermal phenomena and promotes integrated design processes with the aim to support healthier and more durable constructions. Technological advances in materials composition, building components and construction processes have meant that the physical response to moisture is less predictable, increasing the need for accurate building physics models. Modern tools are hence compared, offering comparative insights. Widely used hygrothermal simulation software is compared, showing the relationships among inputs, assumptions and output validity. Moreover, the book explains how relevant policies, codes and international standard adopt different hygrothermal models to assess moisture and condensation risk and provides a review of the current regulatory state of the art in regard to moisture and condensation risks.

In brief, Brambilla and Sangiorgio offer their deep understanding of the topic to their readers and showcase the importance of executing a comprehensive design and plan of building practices to avoid moisture accumulation in building envelopes.

This timely book will help both newcomers and experts, as it candidates to be the simple-to-use and still complete reference that deals with the topic of *Moisture and Buildings*.

Umberto Berardi
Canada Research Chair in Building Science, Ryerson University, Toronto, ON, Canada

Moisture and buildings

1

In 2015, members of the United Nations (UN) signed an Agenda for Sustainable Development, which included a 15-year plan to provide a more sustainable future for all by achieving the 17 Sustainable Development Goals (SDG) (Sustainable Development Goals, 2016). These goals address global challenges such as climate change, health and well-being, which are of direct relevance to the construction sector. UN-SDG 13 *Climate action* and UN-SDG 11 *Sustainable cities and communities* represent a move towards increasingly ambitious energy and environmental performance requirements in the way we design and construct buildings. The resulting pressure is driving the development of new construction methods, practices and technologies, with a corresponding increase in innovative building components and materials. At the same time, UN-SDG 3 *Good health and well-being* calls for tighter indoor air quality (IAQ) standards which, given the increasing density of the urban environment and the need to stem the alarming trend in levels of air pollution, results in a reduced tendency of building occupants to open their windows and ventilate the internal spaces. Further, as recent events have demonstrated, global warming is significantly expanding the number of bushfire-prone areas, thereby exposing a growing number of large cities to smoke hazards and contributing to a further decrease in the external air quality. Also for these reasons, recent building sustainability standards have adopted more stringent envelope airtightness requirements.

The combination of new construction methodologies, stricter airtightness requirements and the changing social and cultural context that influences the way we live inside buildings has created unprecedented challenges for the built environment. In modifying indoor and outdoor environments and the building envelopes that serve as a filter between the two, we are changing the physical parameters of the ways in which buildings behave and respond to climatic stimuli. In particular, increased importance is being directed towards hygrothermal analysis for both new constructions and remedial works. The major issues with moisture-related failures in buildings are the time it takes for negative effects to become visible, which usually occurs only after a period of occupation, and in most cases, the absence of practical and easy-to-adopt solutions. The lack of clear roles and responsibilities in the current design and construction process also poses a risk in regard to moisture management. Yet the changed building requirements have a significant impact on their hygrothermal behaviour—greater airtightness often translates into higher indoor humidity; new bio-based materials are more prone to organic proliferation; and different construction methodologies influence heat and moisture transfer through the envelope.

Understanding and predicting the way in which buildings and moisture may interact should be an important step in the design process, aiming to minimise

Moisture and Buildings. DOI: https://doi.org/10.1016/B978-0-12-821097-0.00005-9

possible negative long-term consequences. Moisture-related issues never have a simple solution, since they involve multiple factors, including design, construction, maintenance, materials, climate and occupation pattern. Thus, while the topic is attracting growing attention among researchers, designers and practitioners, the pace with which actual change is occurring is still too slow.

This book provides a critical overview of current research, knowledge and policy frameworks, and presents a comprehensive analysis of the implications of moisture and the importance of accounting for it during the design process. It responds to the urgent need for a systematic organisation of the existing knowledge to identify research gaps and provide directions for future development. The ultimate goal is to increase awareness of the multifaceted implications of hygrothermal phenomena and promote integrated design processes that lead to healthier and more durable constructions.

1.1 Moisture-related risk and trends in construction and design

For decades, thermal response has been the main priority in building performance design and assessment, with specific focus on insulation from colder outside environments. This focus is reflected in building policy and design and construction practices. In general, one in three homes suffers from excessive dampness and mould proliferation (WHO, 2009), which is exacerbated by inadequate architectural strategies, poor construction practices and inadequate maintenance, resulting from a lack of awareness and knowledge of moisture management in the construction sector (Brambilla & Sangiorgio, 2020). Building sustainability standards, emerging construction trends and changing occupancy patterns are modifying the hygrothermal behaviour of buildings, further aggravating moisture-related issues.

1.1.1 Airtightness

The building envelope is a membrane that separates the indoor and outdoor environments, acting as a regulator between the two. However, the degree of permeability of this membrane has changed over time. With the invention of heating, ventilation and air conditioning (HVAC) systems, the indoor environment has been increasingly detached from the outdoor, as HVACs offer a means of independently controlling and managing indoor spaces. Although the energy crisis prompted a reconsideration of the value of passive design—that is, using the outdoor environment to achieve the desired indoor conditions naturally, thereby reducing energy consumption—the trend towards blocking heat, air and moisture (HAM) transfer through the building envelope continues. A clear example is the meticulous attention that most energy standards pay to the envelope's airtightness. This is mainly due to the dual necessities of energy conservation and protection. The latter, in particular, is becoming an increasingly important concern in the construction sector worldwide. Indeed, the increase in outdoor pollution levels, due to urbanisation and associated human activity, bushfires, and the latest pandemic, is leading to calls for sealed envelopes that can protect the occupants' health by completely blocking air exchange and increasing the reliance on mechanical ventilation and air filtering.

However, higher levels of airtightness are associated with higher indoor moisture loads, which is a risk factor in condensation and mould growth management (Brambilla & Sangiorgio, 2020). The reduction in air leakage minimises the uncontrolled exchange with the exterior, resulting in a system that greatly relies on the users to either correctly engage with the ventilation system or naturally ventilate the indoor spaces to dissipate the excess latent loads. Nevertheless, there is evidence that designers are overestimating the actual air change rates, as occupants tend to under-ventilate the indoor environment, further contributing to the increased occurrence of moisture-related problems. The importance of a moisture-aware design is also highlighted by the unpredictability of the building occupants' behaviour. Better understanding hygrothermal processes and the risks associated with design choices can help to increase the building's resilience to moisture-related failures.

1.1.2 Construction methodologies

In the architectural process, the choice of materials is usually driven by multiple factors, such as architectural intent, energy efficiency and thermal behaviour, availability of the product, code requirements and cost. Emerging construction practice and sustainability standards are, however, driving the market and promoting the use of bio-based and often organic products, such as timber, straw and sheep's wool, among others. These materials are particularly sensitive to moisture, which often reduces their thermal performance and durability due to mould growth, as they provide high concentrations of organic compounds that can be easily digested by fungi (Hoang, Kinney, Corsi, & Szaniszlo, 2010).

Another trend in facades design and construction is the increased use of materials with low vapour permeability. A clear example is the widespread use of metal claddings, which already constitute a vapour barrier for the envelope, preventing moisture exchanges between indoors and outdoors. Additionally, there is no clarity around the responsibility of builders during the construction process for preventing moisture-related durability issues. It is of utmost importance that materials are stored correctly during construction and protected from the weather, as rain may cause an additional initial moisture load which, following construction, is difficult to dry, especially in new highly performing and airtight envelopes (Geving & Holme, 2010). This is further aggravated by the limitations imposed by the building codes for fire safety which, depending on the code and the building typology, can discourage the use of permeable membranes due to issues of combustibility, favour the use of metallic panels as weather barriers, or limit the continuity of the ventilated cavity behind the cladding in order to prevent rapid spread of smoke or flames. A metal cladding with no ventilated cavity behind it poses a clear risk of condensation, deterioration and long-term damage to the envelope.

The higher risk of moisture-related insurgence does not, however, depend on the choice of construction practice or material itself, but reflects the fact that the policy requirements and design process are not updated accordingly. On one hand, there is a push towards a different means of construction but, on the other, the process is not able to fully capture the new issues arising from it. Increasing awareness of the physics and principles of the hygrothermal process and the implications for buildings and occupants is a first step towards a moisture-safe design.

1.1.3 Occupancy patterns

Between home and office, humans spend up to 90% of their time indoors (Klepeis et al., 2001), with considerable impact on the indoor environment. The significant amount of time spent indoors translates into a demand for higher comfort to ensure healthy spaces. Building efficiency standards and construction codes are moving towards stricter requirements for more accurate and stable temperatures, resulting in an increased reliance on mechanical air conditioning and ventilation systems that are often unable to manage the indoor latent loads. On the other hand, new buildings with augmented thermal protection tend to shift towards an internal-dominated mode, meaning that the occupancy patterns become the drivers of the quality of the indoor environment. Occupancy patterns identify those actions and activities that building occupants perform indoors and which have an impact on the building's hygrothermal behaviour.

Showering, cooking, drying laundry inside, breathing and, as COVID-19 has required, performing high intensity physical activities (such as working out) release a significant amount of moisture, increasing the indoor humidity. Given the new ways of smart-working that 2020 has shown to be possible, these trends are only likely to increase in the future. However, these additional and uncontrolled sources of moisture, coupled with reduced ventilation rates, may lead to moisture-related issues, in the form of both condensation failures due to higher vapour pressure on the envelope and indoor mould growth due to the higher water availability on the indoor surfaces. Minimising the activities that increase indoor moisture loads and engaging in those that reduce it, such as ventilating indoor spaces, is the responsibility of the building's occupants. However, it is important to design buildings capable of providing a certain degree of resiliency to increasing levels of indoor moisture generation. This is only possible by taking account of hygrothermal processes from the early design stage and understanding and managing the possible implications of high moisture levels. Finally, raising awareness and promoting mould risk-conscious behaviour through social education is essential (Australian Building Code Board, 2019).

1.2 Dampness-related impacts on humans

A damp environment usually implies a higher risk of mould growth, which has significant negative implications for human health. Indoor mould has been related to adverse health symptoms such as severe allergic asthma, hypersensitivity pneumonitis and allergic alveolitis, allergic rhinitis and sinusitis (Brambilla & Sangiorgio, 2020). Mould is responsible for early biodeterioration of building materials, requiring anticipated renovation works. The economic impact of mould has not been investigated in all countries. In 2001, however, it was estimated that Germany suffered an economic loss of more than $200 million due to mouldy indoors, and there is reason to believe that the magnitude may be at least comparable elsewhere. The issue is magnified by the difficulty of detecting mould before it is fully germinated. Indeed, the visible manifestation occurs only at the final proliferation stage, which usually starts within the building envelope.

Unfortunately, damp indoor spaces are not associated only with old constructions. Some studies investigating the increased occurrence of adverse health symptoms, such as asthma, rhinitis and hay-fever, in newly built energy-efficient buildings have identified mould as one of the primary causes. A cross-sectional study conducted in China on more than 7000 children correlated newly-built homes with rising cases of asthma and allergies; the use of air- conditioners accounted for 7%–17% of rhinitis and eczema cases (Sun et al., 2019). Similar results linking central air-conditioning systems with an increased incidence of asthma have also been reported in Germany (Jacob et al., 2002), and in the UK (Sharpe, Thornton, Nikolaou, & Osborne, 2015). These results have been notionally linked to the re-circulation of indoor air with very limited exchange with outdoor fresh air, which can lead to high concentrations of indoor fungal spores and chemical and physical contaminants.

Newly built houses have also been associated with sick building syndrome (SBS), which is a constellation of symptoms ranging from irritation of the eyes, nose and throat, headache and general fatigue (Finnegan, Pickering, & Burge, 1984). Although lethargy and headache are found to be the most common symptoms (Mendell & Smith, 1990), 9% of cases can result in respiratory difficulties. SBS is usually caused by indoor pollution, inadequate temperature and humidity, and the presence of mould. However, it is not always possible to identify a specific causal factor; all these parameters are likely to be implicated (Godish, 1994).

1.3 Book structure

This book provides an overview and critical assessment of current knowledge of the interactions between moisture and buildings. It does not focus on a comprehensive catalogue of moisture-safe construction details, but provides the knowledge necessary to adopt a moisture-aware design approach and implement strategies aimed at minimising the risks that result from uncontrolled or underestimated moisture implications. It delves into the different interactions between materials, climate, construction techniques and occupancy patterns that determine a building's hygrothermal performance.

Chapter 2, Principles of hygrothermal processes, presents the theoretical background to hygrothermal processes, explaining the physical consequences that can be observed at the macro level. This chapter also looks at the micro-scale, defining the laws that determine HAM transfer in buildings. It does not elaborate on the analytical approaches to this topic as the literature contains numerous HAM models based on different interpretations of the same physical phenomenon. No one model is more 'correct' than another, and different models may suit different applications. For this reason, the chapter focuses on the physical description to provide the reader with a general understanding of the main drivers and the most important parameters. The core of the book is contained in the three following chapters, which focus on the three main consequences of the interactions between moisture and buildings: condensation due to hygrothermal transfer within the envelope; mould growth due to moisture availability and the buffering of indoor moisture by the building materials.

Chapter 3, Durability, condensation assessment and prevention, presents a comprehensive examination of the durability issue. It discusses condensation as a consequence of moisture transfer through the building envelope by considering the application of hygrothermal principles. The chapter investigates the causes and effects of condensation from the perspective of design and construction, showing how the design approach may be the vital factor in determining whether a building will suffer from durability issues. Prevention strategies are considered, especially in regard to tools and approaches that are helpful during the design stage. Traditionally, publications and research in this field have focused on cold continental climates, mainly due to the high relevance of thermal bridges in moisture-related issues. However, with the broad diffusion of air conditioning systems, condensation is also emerging as a problem in warm climates, where context, construction techniques and requirements are completely different. This chapter explores this difference and discusses the challenges and opportunities in both cases. Several case studies and possible solutions are discussed.

Chapter 4, Health and mould growth, deals with biodegradation and biological germination in buildings. It analyses mould growth in buildings as a consequence of high water availability within the building envelope. It discusses the different types of fungal species that are commonly found indoors, the parameters that influence their germination and growth, and the implications for both building materials and building occupants' health. It also presents an overview of the growth models available for assessment, as well as the state of the art for remediation. Mould growth is an emerging field of research, and this chapter comprehensively reviews the current state of knowledge, identifying the gaps that need to be filled to identify a reliable strategy for mould prevention. Several case studies of how fungi and bacteria may be used to improve construction techniques by taking advantage of their biological characteristics are also discussed.

Chapter 5, Moisture buffering of building materials, introduces the concept of moisture buffering and investigates the effects on indoor environmental quality. All building materials have the capacity to store and release moisture, interacting with the surrounding environments. Highly hygroscopic materials in contact with indoor air can therefore play a major role in balancing the indoor relative humidity. This chapter discusses the moisture uptake capacity of common building materials, analysing the potential effects on both building energy consumption and occupants' thermal comfort. It also presents some case studies where this physical phenomenon has been used to control the internal environment while integrating artistic features into the interior lining. These three main implications of moisture interaction with buildings can be assessed or predicted at the design stage with the aid of software. Several hygrothermal transient tools have been developed based on the different HAM models.

Chapter 6, Hygrothermal modelling, discusses the similarities and differences among the most common models. It explains the fundamental assumptions common to all the software as well as the nature, potentialities and possible uses of each This chapter differs from other texts on the topic as it does not focus on the mathematical model behind the software but, rather, discusses the pros and cons of using them during the design stage, including their ease of use and potentialities. Although numerous sophisticated software programs are available to predict hygrothermal processes with a high degree of accuracy, mould growth and condensation still affect up to 50% of the building stock.

Chapter 7, Building codes and standards, investigates the policy framework and existing regulations around hygrothermal assessment. It analyses the main differences between the American, European and Australian building codes to provide an understanding of current legislation. This chapter identifies the major shortcomings of the relevant codes, as well as their potentialities and suggests future directions for research to promote a moisture-aware design process.

Finally, Glossary presents a handy syllabus that can be used to navigate this book and to develop a deeper understanding of some of the topics. It contains a list of relevant terminology, concepts and references.

References

Australian Building Code Board. (2019). *Handbook: condensation in building.*

Brambilla, A., & Sangiorgio, A. (2020). Mould growth in energy efficient buildings: causes, health implications and strategies to mitigate the risk. *Renewable and Sustainable Energy Reviews*, 110093. Available from https://doi.org/10.1016/j.rser.2020.110093.

Finnegan, M. J., Pickering, C. A. C., & Burge, P. S. (1984). The sick building syndrome: prevalence studies. *British Medical Journal, 289*(6458), 1573–1575. Available from https://doi.org/10.1136/bmj.289.6458.1573.

Geving, S., & Holme, J. (2010). The drying potential and risk for mold growth in compact wood frame roofs with built-in moisture. *Journal of Building Physics, 33*(3), 249–269. Available from https://doi.org/10.1177/1744259109351441.

Godish, T. (1994). *Sick buildings: definition, diagnosis and mitigation.*

Hoang, C. P., Kinney, K. A., Corsi, R. L., & Szaniszlo, P. J. (2010). Resistance of green building materials to fungal growth. *International Biodeterioration and Biodegradation, 64*(2), 104–113. Available from https://doi.org/10.1016/j.ibiod.2009.11.001.

Jacob, B., Ritz, B., Gehring, U., Koch, A., Bischof, W., Wichmann, H., & Heinrich, J. (2002). Indoor exposure to molds and allergic sensitization. *Environmental Health Perspectives*, 647–653. Available from https://doi.org/10.1289/ehp.02110647.

Klepeis, N. E., Nelson, W. C., Ott, W. R., Robinson, J. P., Tsang, A. M., Switzer, P., ... Engelmann, W. H. (2001). The national human activity pattern survey (NHAPS): a resource for assessing exposure to environmental pollutants. *Journal of Exposure Science & Environmental Epidemiology*, 231–252. Available from https://doi.org/10.1038/sj.jea.7500165.

Mendell, M. J., & Smith, A. H. (1990). Consistent pattern of elevated symptoms in air-conditioned office buildings: a reanalysis of epidemiologic studies. *American Journal of Public Health, 80*(10), 1193–1199. Available from https://doi.org/10.2105/AJPH.80.10.1193.

Sharpe, R. A., Thornton, C. R., Nikolaou, V., & Osborne, N. J. (2015). Higher energy efficient homes are associated with increased risk of doctor diagnosed asthma in a UK subpopulation. *Environment International, 75*, 234–244. Available from https://doi.org/10.1016/j.envint.2014.11.017.

Sun, Y., Hou, J., Sheng, Y., Kong, X., Weschler, L. B., & Sundell, J. (2019). Modern life makes children allergic. A cross-sectional study: associations of home environment and lifestyles with asthma and allergy among children in Tianjin region, China. *International Archives of Occupational and Environmental Health*, 587–598. Available from https://doi.org/10.1007/s00420-018-1395-3.

Sustainable Development Goals. *Department of Economic and Social Affairs*. (2016).

WHO. (2009). *WHO Guidelines for Indoor Air Quality: Dampness and Mould*. World Health Organization. Available from https://www.euro.who.int/__data/assets/pdf_file/0017/43325/E92645.pdf?ua = 1.

Principles of hygrothermal processes

2

The hygrothermal performance of a building is determined by the interdependent sorption, desorption and storage of heat and moisture within the building envelope. Yet, moisture has long been overlooked in building design, in which the main focus has been on indoor temperature, heating, cooling and thermal properties of the envelope. This approach has resulted in an imbalance in applied building physics, where recognition of the importance of transient thermal assessments in fully understanding building behaviour often drives energy efficient design, while moisture is only considered in relation to condensation assessment. Indeed, hygrothermal performance assessment is seen as part of the final validation of a design and is often delegated to external consultants. This overlooks the importance of moisture as a design parameter that impacts the envelope's durability, the building's energy efficiency and occupants' health and comfort. This superficial approach has led to significant failures in the building sector, with numerous examples worldwide of the destructive effects of design that has failed to take moisture into account. A sound understanding of hygrothermal principles and processes is necessary to avoid these issues by adopting a moisture-aware approach to building design, construction, operation and maintenance.

This chapter presents an up-to-date overview of heat, air, moisture (HAM) transfer and hygrothermal principles. It first discusses moisture transport and storage within the building envelope, which is a necessary starting point for understanding the complex integration of heat, moisture and air transport in the built environment. It then explains the physical principles of psychrometry. The chapter scaffolds the knowledge necessary to understand the following chapters and the intercorrelations between moisture and buildings.

2.1 Moisture in buildings

Moisture is a term used to define the presence of small amounts of water. In buildings, moisture can be found in all three states of water: solid, liquid and gas. Accordingly, the term *moisture transfer* refers to all movements of liquid and vapour across the building envelope and the whole building. Liquid transport refers to the movement of clusters of water molecules through diffusion, gravity, pressure-induced flow and capillary suction. Vapour transport indicates the movement of single water molecules, which can occur through diffusion or convection. The former is driven by differences in vapour concentration (from more to less) and temperature (from hot to cold) between two zones; it is generally a slow

Moisture and Buildings. DOI: https://doi.org/10.1016/B978-0-12-821097-0.00007-2

process and may lead to significant accumulation of moisture within the envelope, thus presenting a risk of subsequent deterioration of materials. The latter refers to the condition in which water molecules are carried by air and migrate across the building envelope through convective flow, driven by a difference of air pressure (from high to low). This transport can take place through both controlled air flows, such as ventilation sources and mechanical ventilation systems, or uncontrolled ones, such as holes, cracks and air infiltration. Air infiltration is one of the last moisture transfer processes to have been acknowledged, and most of the current simulation software does not consider it at all or implements it in a simplified way.

Ideally, in the absence of other materials, the thermodynamics of moisture is defined by the unique correlation between temperature and vapour pressure at saturation, which is the maximum vapour pressure that moisture can establish, and beyond which vapour condenses into water. This unique correlation is the basis of simplified hygrothermal assessments and psychrometric calculations.

2.1.1 Physics of moist air

Air is a physical mixture of various gases and water vapour. It is composed of nitrogen (78%), oxygen (21%) and 1% of carbon dioxide and inert gases, such as hydrogen, helium, neon, argon and water vapour. Air without any moisture content is called dry air, but this is a pure physical quantity as completely dry air does not exist, whereas the moisture content of air varies according to geographical and atmospheric conditions in the range of 1%–3% by mass. However, dry air is a useful concept for describing the properties of air and it is widely used in Psychrometry, the science that studies the physics of gas–vapour mixtures, including moisture content, humidity and temperature. Psychrometric calculations deal with the physical processes and hygrothermal behaviour of moist air. Thus, they can be used in building science for a wide range of applications, especially for mechanical heating, ventilation and air conditioning (HVAC) system design. This approach relies on a steady-state simplified model, which assumes low moisture content, constant material properties and constant boundary conditions.

Psychrometry's basic assumption is that there is a unique relationship between temperature and saturation vapour pressure, resulting from the ideal gas law:

$$p \cdot V = n \cdot R \cdot T \tag{2.1}$$

where p is the air pressure, V the volume occupied, n is the amount of gas, or mole, T is the temperature, and R is the universal gas constant, equal to 8.31446 J/(mol K).

At any given temperature it is possible to define an equilibrium between the liquid, solid and vapour state of water. However, air can absorb and carry moisture until it reaches a certain value, called the saturation point, beyond which the additional moisture is transformed into dew. The dependence of saturation vapour pressure on temperature is almost an exponential curve, and can be calculated through the mathematical equation:

$$p_{sat} = p_{c,sat} \cdot exp\left[2.3026 \cdot k\left(1 - \frac{T_c}{T}\right)\right] \tag{2.2}$$

with T_c representing the temperature at the triple point above which water exists as vapour only, and $p_{c,sat}$ the related saturation pressure. The parameter k depends on the temperature in K:

$$k = 4.39553 - 6.2442 \cdot \left(\frac{T}{1000}\right) + 9.953 \cdot \left(\frac{T}{1000}\right)^2 - 5.151\left(\frac{T}{1000}\right)^3 \tag{2.3}$$

The temperature at which vapour pressure reaches the saturation value is called the *dew point*. Because air is a mixture of different gases among which no chemical reactions occur, Dalton's law of partial pressure for gases is applicable. This law states that the total pressure of a mixture of gases is equal to the sum of the partial pressure of each gas. Considering atmospheric air as the sum of dry air and water vapour, it is possible to state that the atmospheric pressure exerted by the air is the sum of the partial pressures of dry air and water vapour:

$$p_{tot} = p_{vap} + p_{air} \tag{2.4}$$

The higher the content of moisture within the air, the higher is its partial vapour pressure. The saturation point indicates that state in which the partial vapour pressure reaches its maximum before forming dew. Thus, at saturation, the total pressure can be described as:

$$p_{tot} = p_{sat} + p_{air} \tag{2.5}$$

When the partial vapour pressure is equal to the saturation vapour pressure, the moist air is called saturated air. In the built environment, the dry air vapour pressure is the atmospheric value (1 atm at sea level), hence, the total saturation pressure depends uniquely on the temperature, as per Formula 2.1. Higher saturation vapour pressure can be expected at higher temperatures.

In psychrometry, the point at which air is considered saturated or non-saturated is defined by the ideal gas law (Formula 2.1). In building applications, the partial vapour pressure and temperature are used to derive a number of parameters used for HVAC system calculation, simplified condensation assessments and thermal comfort analysis, as described in subsequent chapters.

2.1.1.1 Humidity

In psychrometry, several factors related to the quantity of vapour contained in the air are correlated with one another and are used for different applications. Usually, the state of air is somewhere in the range from dry to saturated, and the parameter that expresses the ratio between vapour pressure and vapour saturation pressure at any given temperature is called relative humidity (RH):

$$RH = 100\frac{p_v}{p_{sat}} \quad (2.6)$$

This value is a dimensionless number, usually expressed in the range 0 to 1 in most building applications, or as a percentage (0% to 100%), where 1 represents the state of saturation (when the partial vapour pressure is equal to the saturation vapour pressure). When moist air is subjected to a change of temperature, it is possible to define the new relative humidity through a simple equation, based on the unique correlation between temperature and saturation vapour pressure:

$$RH_2 = RH_1 \cdot \left(\frac{p_{sat1}}{p_{sat2}}\right) \quad (2.7)$$

Hence, when cooling a building, the relative humidity will increase accordingly, while heating is associated with a reduction of relative humidity.

Other parameters can be used to express the presence of vapour in moist air. These do not substitute for relative humidity, but can be used to complete the description of the state of the moist air. The first is the absolute humidity, defined as the mass of vapour per unit volume in the sample of air under any given condition. The second is the humidity ratio, which represents the ratio of the actual mass of vapour in the air to the mass of dry air and is expressed in kilogram water/kg dry.

2.1.1.2 Thermal state of moist air

To give a complete overview of the parameters that describe the state of moist air, it is important to recall the definition of the different temperatures that express the thermal aspects of air. As with humidity, several parameters can be identified:

- The dry-bulb temperature is the temperature normally measured by thermometers; it is therefore the ordinary atmospheric temperature. The absence of a prefix specifying the type of temperature usually means dry-bulb temperature.
- The wet-bulb temperature (WBT) can be approximated to the temperature monitored by a thermometer covered with a wet cloth. The wet-bulb temperature at saturation corresponds to the atmospheric temperature, otherwise it is lower than the dry-bulb temperature; the difference between the two is usually referred to as wet-bulb depression.
- The dew point temperature is the temperature at which vapour starts to condense into dew, that is, air is in its saturated condition.

2.1.1.3 Heating and cooling of moist air

Air can be heated and cooled. Psychrometric calculations assess the state of the air before and after hygrothermal processes, including heating and cooling. When the air temperature increases, its capacity to carry moisture increases, so the relative humidity decreases. The change of partial pressure is directly proportional to the temperature, as per the ideal gas law (Formula 2.1). However, according to the vapour pressure equation (Formula 2.2), the change in saturation pressure is exponential, resulting in a greater change. During the change of temperature, the total heat absorbed by the air is called enthalpy, which includes both latent and sensible heat.

Latent heat is the heat absorbed or released during the change of phase of water. Water vapour carried inside the air is subjected to transformations during a change in temperature of the air, thus, the latent heat is related to the amount of moisture present in the air. It is possible to calculate the latent heat as:

$$Q_l = m \cdot L \quad (2.8)$$

where m is the mass involved and L the specific latent heat. The specific latent heat is defined as the amount of heat necessary to complete a phase change of a unit of mass of a substance, which is a property of the substance independent from its size and is usually tabulated.

Sensible heat is related to the mass of dry air not subjected to phase change, as no vapour is present. When the temperature changes, the heat absorbed or removed can be calculated as:

$$Q = m \cdot 0.133 \cdot \Delta T \quad (2.9)$$

where m is the mass, ΔT is the temperature gradient and 0.133 is the specific heat of air expressed in kcal/kg.

The total heat of air can be calculated as the sum of these two components. Considering that latent heat depends on the dew point temperature and sensible heat on the dry-bulb temperature, the total quantity of thermal energy in the air depends on both these temperatures.

2.1.2 The psychrometric chart

The psychrometric chart (Fig. 2.1) graphically approximates the properties of moist air.

It can be used to solve calculations and processes involving the above parameters. This type of chart expresses the thermodynamic parameters of moist air at a constant atmospheric pressure as follows:

- Dew point curve is the last curve of the graph; it represents the point of condensation of vapour and is equal to the fully saturated temperature.
- Dry-bulb temperature is plotted on the abscissa.
- Wet-bulb temperature is shown on sloped lines where the inclination is the heat of vaporisation required for the water to saturate the air at a given relative humidity; these lines intersect the dew point curve at the dry temperature point.

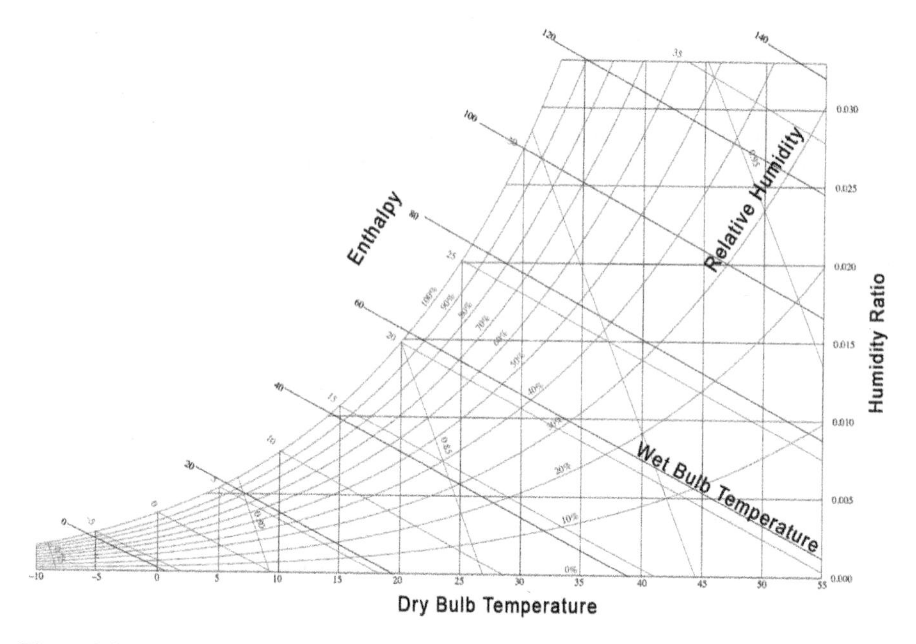

Figure 2.1 The psychrometric chart described the physical and thermodynamic properties of moist air.

- Relative humidity is shown in curves. The dew point curve is equal to 100% of relative humidity.
- Humidity ratio is plotted on the ordinate.
- Specific enthalpy is shown on sloped lines; lines of constant enthalpy are parallel to lines of constant WBT.

When two of these parameters are known, it is possible to determine the others.

2.1.2.1 *Moving on the chart*

Air can be heated without change in the vapour concentration through sensible heat. In this case, the movement on the chart would be horizontal when moisture content humidity ratio and dew point remain constant. In contrast, when latent heat is removed or added, the movement on the chart would occur on the vertical lines, with dry bulb temperature constant. However, most building applications include a mixed heating and cooling process that involve both sensible and latent heat. Thus, the movement on the psychrometric chart occurs on the oblique lines, depending on the initial and final states. The various transformations to which air may be subject can lead to meet the dew point line, indicating that the vapour present in the air will form dew. In relation to building applications, this means moisture is forming condensation. It usually represents a situation in which the air is cooled but not de-humified. Chapter 3, Durability, condensation assessment and prevention, discusses in more detail how the psychrometric chart can be used for simplified condensation assessment during the design phase.

It must be noted that psychrometric calculations rely on the unique correspondence between vapour and temperature found in moist air. However, this does not apply within building envelopes, where porous materials are inhomogeneous and anisotropic, and their pore system cannot be described as a hydraulic network. Indeed, in a porous medium, the transport laws only represent an approximation, and the only possible approach is phenomenological (Hens, 2017).

2.2 Vapour in open porous materials

Building materials are generally porous, with moisture filling the pores in its three different states: gas, liquid and solid. The parameters that describe this condition of the materials are density (*d*), which is the mass per unit volume of material, and porosity (*Y*), the volume of pores per unit volume of material expressed in %m^3/m^3. The presence of air is described by four different parameters:

- Air content (w_a): mass of air per unit volume of material (kg/m^3).
- Air ratio (χ_a): mass of air per unit mass of dry material (%kg/kg).
- Volumetric air ratio (Ψ_a): volume of air per unit volume of material (%m^3/m^3).
- Air saturation degree (S_a): ratio between the air content and the maximum admissible (%), also depictable as the percentage of pores filled with air in relation to all those accessible.

The same parameters can be defined to describe the moisture filling the pores:

- Moisture content (w): mass of moisture per unit volume of material (kg/m^3).
- Moisture ratio (χ): mass of moisture per unit mass of dry material (%kg/kg).
- Volumetric moisture ratio (Ψ): volume of moisture per unit volume of material (%m^3/m^3).
- Moisture saturation degree (S): ratio between the moisture content and the maximum admissible (%), also depictable as the percentage of pores filled with moisture in relation to all those accessible.

If no other substances are present in the pores, the sum of air and moisture saturation degrees must be equal to 1. Depending on the quantity of air or liquid present in the material, it is possible to define several degrees of saturation, ranging from 0%, when only air is present and representing a dry material, and 100%, when only liquid is present and representing a wet material. Nonetheless, in reality a material can present a range of values between the extremes. Fig. 2.2 represents the range of different proportions that can be found between the moisture and air content within a porous structure.

2.2.1 Sorption and desorption isotherm

Under standard atmospheric conditions, the pores are partly filled with liquid and partly with moist air. Hence, the standard psychrometric laws for moist air should

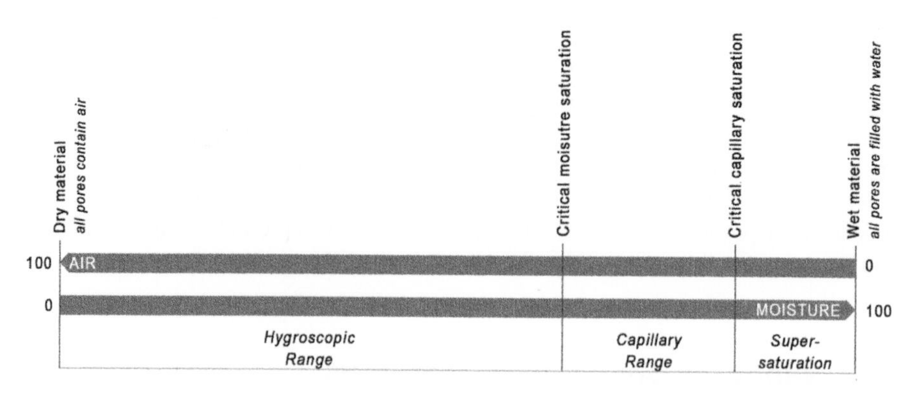

Figure 2.2 Range of different proportions that can be found between the moisture and air content winthin a porous material.

apply and indicate a linear relationship between the moisture content in the pores and the air relative humidity. However, this behaviour is not observed in reality, and materials do not display a linear relation between the moisture content of the pores and the air RH. The reason must be searched in the materials' tendency to absorb water to a greater or lesser extent, depending on the hygroscopic nature of the material itself. For example, timber is more hygroscopic than metal and tends to absorb more moisture when exposed to a humid environment. This ability of hygroscopic materials is also referred to as sorption, which is the sum of two processes: absorption and adsorption. Absorption indicates a bulk phenomenon in which a fluid is dissolved by a liquid or a solid (absorbent), while adsorption is a surface phenomenon where atoms or molecules from gas or liquid adhere to a surface.

When referring to moisture transfer in building materials, sorption encompasses not only the external surfaces, but also the internal surface defined by the pores. Clearly, the gross surface is much less extensive than the surface offered by the pores; hence the intrinsic sorption behaviour of porous materials can involve a significant amount of moisture. Materials exposed to a humid environment store moisture through adsoprtion processes, and the curve that describes the relation between RH and moisture content within a material is called the sorption isotherm. This phenomenon is observed in both directions, with materials storing moisture when exposed to a humid environment and releasing moisture when exposed to a dry environment. This last process is called desorption and generally exceeds sorption, with a higher moisture content attained during the drying of the pores. Thus, the curve that can be traced during the desorption process from saturation to dry, does not perfectly overlap the sorption curve, hence creating a hysteresis. The sum of the two curves is called the sorption isotherm (Fig. 2.3).

The sorption curve identifies two different regions: the hygroscopic range, usually up to RH 98%, and the capillary range. Empirically, the sorption curve can be

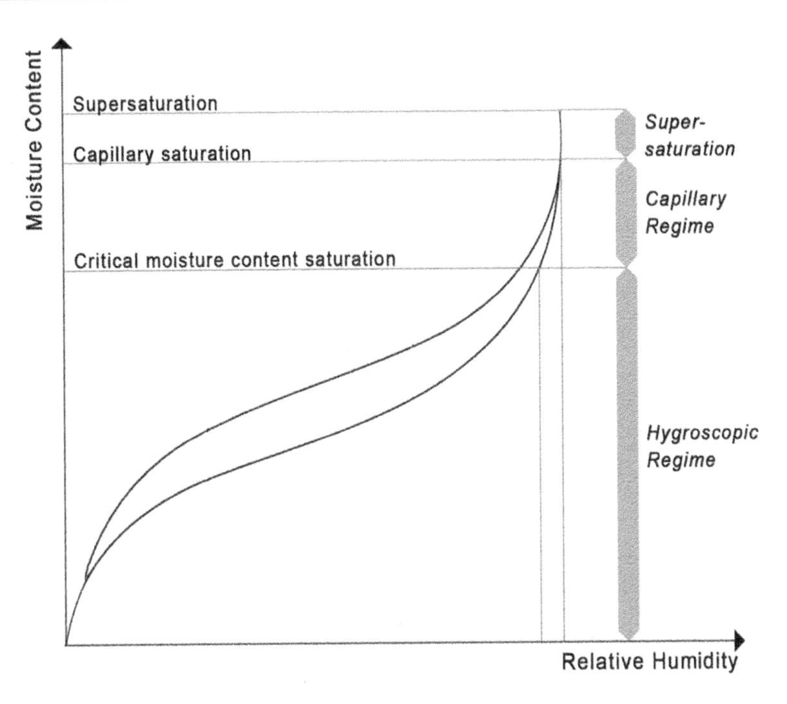

Figure 2.3 The curve represents the moisture sorption and desorption of porous building materials, identifying the hygroscopic and capillary regions defined by that saturation state within the pores.

determined by following the ASTM Standard C1498, which prescribes a pressure plate measurement process (ASTM C1498-04a, 2010). There are also several mathematical expressions in the literature, but none is universally accepted. The following equation has been successfully used to determine the sorption curve of various porous materials when RH is below 95% (Hens, 2017):

$$w_H = \frac{RH}{a_H \cdot RH^2 + b_H \cdot RH + c_H} \tag{2.10}$$

where the coefficient a_H, b_H and c_H are specific to each material and different for sorption and desorption.

At relative humidity below 40%, the pores contain mainly water in the form of gas and, thus, the governing moisture transport process is adsorption, with water molecules diffusing into the pores and adhering to the surface. At higher relative humidity, the pores are filled with liquid and the governing process is capillary suction. However, the distinction is not necessarily clear-cut; moisture diffusion and capillary suction may coexist at the boundary of the two regions, where vapour diffuses in large pores and condenses in smaller ones, while water evaporates from small pores into larger pores. The first phase of molecular adsorption involves a monolayer of water molecules. The hygroscopic moisture content (w_H) can be expressed through Langmuir's relation:

$$w_H = \frac{M_w \cdot A_p}{A_w \cdot N} \cdot \frac{Q \cdot RH}{1 + Q \cdot RH} = 2.62 \times 10^{-7} \cdot A_p \frac{Q \cdot RH}{1 + Q \cdot RH} \tag{2.11}$$

where M_w is equal to 0.018016 kg and represents the mass of 1 mole of water, A_w is $11.2\text{x}10^{-20}$ m^2 being the area occupied by a molecule of water, N is the Avogadro number ($6.03\text{x}10^{23}$ molecules/mole), A_p is the surface offered by the pores and Q is the heat exchanged during the adsorption process. The latter can be calculated as:

$$Q = k \cdot exp\left(\frac{(l_a - l_b)}{R \cdot T}\right) \tag{2.12}$$

where k is the adsorption constant, l_a is the heat of adsorption expressed in J/kg, l_b is the heat of evaporation (J/kg) and R is the gas constant of vapour. This indicates that, in the hygroscopic region, the moisture content depends on the heat exchanged and the pore surface. In building applications, this dependency suggests that, at low RH, materials with very small pores are more hygroscopic than materials with a larger pore structure. For example, bricks and limestone both have approximately 33% porosity, but the diameter of the pores in the former is approximately 8 micron, compared to 0.1 micron in the latter, resulting in a pore surface up to 6000 times larger and a much higher level of hygroscopic activity.

At higher RH, the process involves multiple layers of water molecules with a weaker bond which, as they increase, may fill the smallest pores and touch each other, indicating the beginning of capillary condensation. The surface tension breaks the bonds and rearranges the molecules in a more stable water drop with a meniscus at both ends. The process is better represented in Fig. 2.4.

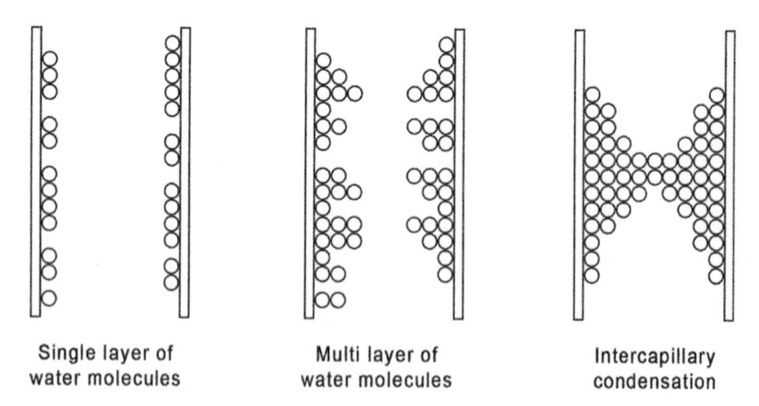

Figure 2.4 The water firstly create a single layer of molecules highly bonded to the surface of the pores. At higher RH, the process involves multiple layers of water molecules that may fill the smallest pores and touch each other. The surface tension breaks the bonds and rearranges the molecules in a more stable water drop with a meniscus at both ends.

The shape of the meniscus also determines the vapour saturation pressure, as a concave meniscus offers higher resistance than a flat one, while a convex meniscus has lower resistance. Formula 2.14 correlates the shape of the meniscus to the saturation pressure, indicating that condensation will occur below RH 100% in the presence of a concave meniscus and above this value when the meniscus is convex. The saturation pressure above a curved meniscus (p'_{sat}) is therefore calculated as:

$$p'_{sat} = p_{sat} \cdot exp\left[-\frac{\sigma_w \cdot \cos(\vartheta)}{p_w \cdot R \cdot T}\left(\frac{1}{r_1} + \frac{1}{r_2}\right)\right] \tag{2.13}$$

where p_{sat} is the saturation pressure above a flat meniscus, p_w and σ_w are the density and surface tension of water, ϑ is the angle formed between the meniscus and the pore wall, and r represents the radii. Assuming that the pore structure is circular, and that the ratio between the p'_{sat} on the curved meniscus to the p_{sat} on a flat one stands for the relative humidity (RH), the formula can be rewritten as:

$$\ln(RH) = -\frac{4\sigma_w \cdot \cos(\vartheta)}{p_w \cdot R \cdot T \cdot d_{eq}} \tag{2.14}$$

This equation indicates that higher pore dimensions require higher RH to achieve capillary condensation, and that the colder it is, the lower is the RH required to start the capillary condensation process. This relation is not valid for RH below 20% and close to 100%. In the former case, the diameter of the pores assumes values below 10^{-9} m, resulting in molecules that do not behave statistically, hence capillary suction theory is no longer valid. In the latter case, the diameter assumes a value near infinity. Further, at RH 100%, all pores should be filled with water, but the air trapped inside the pore structure limits the capillary moisture content. If this situation prevails for years, the air may dissolve into water, shifting the moisture content to saturation.

2.3 Moisture transfer in porous media

At a molecular level, moisture transfer through the envelope is allowed only when the cluster of molecules involved in the process can pass through the pores. Vapour transfer engages molecules separately, while in liquid transfer they travel as clusters of molecules. Given that the molecular diameter of water vapor is 0.28 nm and that clusters have a much larger diameter, some materials allow for vapour transport but not for liquid transport, making them permeable to vapour but impermeable to water. Moisture transfer within the building envelope can occur through different mechanisms, depending on the state of moisture, the driving potential involved and the characteristics of the materials. Table 2.1 summarises the major types of moisture transfer mechanism in buildings.

Airflow is usually admitted in larger pores but is barely possible in micro-pores, while capillary suction is associated with smaller pores. Hence, non-capillary materials, characterised by large pores, do not accept capillary flow, whereas non-hygroscopic materials present pores that are small enough to activate capillarity but

Table 2.1 Possible moisture transfer mechanism for building applications. Which mechanisms intervene will depend on the nature of a material.

Type		Causes
Liquid	Diffusion	Caused by the gradient of liquid concentration between two zones.
	Capillary flow	Differential in capillary suction is related to pore width: the smaller the pore, the higher its suction potential.
	Gravitational flow	Caused by the weight and height of the water molecules; usually occurs in larger pores with no capillary suction.
	Poiseuille flow	Created between two infinitely long parallel plates, separated by a distance with a constant pressure gradient is applied in the direction of flow
Vapour	Diffusion	Caused by differential vapour pressure between the air contained by the pores and the ambient air at either side of the envelope.
	Pressure flow	Caused by the gradient in air pressure at the two sides of the envelope; vapour migrates with air across the material.

too large to collect sorption moisture, while capillary and hygroscopic materials have the finest pores and therefore allow both capillary suction and moisture sorption.

2.3.1 General transport equation

The general transport equation can be expressed as:

$$J_B = -k \cdot \nabla \Phi_B \tag{2.15}$$

where Φ_B is the driving potential and k is the transport coefficient, a characteristic of the medium in which the transport is taking place. J_B is usually defined as the quantity transported across a unit area normal to the direction of the transport; thus, this flux can have three components based on the Cartesian coordinates. The total flux can then be expressed through a system of three different equations where b is the reference axis.

$$\begin{cases} J_{Bx} = -k \cdot \left(\dfrac{d}{d_x}\right) \cdot \Phi_{Bx} \\ J_{By} = -k \cdot \left(\dfrac{d}{d_y}\right) \cdot \Phi_{By} \\ J_{Bz} = -k \cdot \left(\dfrac{d}{d_z}\right) \cdot \Phi_{Bz} \end{cases} \tag{2.16}$$

This set of equations allows different directions of the flux to be defined, acknowledging the anisotropy of building materials, and underpins the different transfer models. Note that the negative sign on the equations describes the movement from a high potential to a low potential.

2.3.1.1 Vapour transfer

Vapour can be transferred through convection and diffusion; hence, the total vapour flow is the sum of these two processes:

$$J_v = J_{conv} + J_{diff} \tag{2.17}$$

Diffusion, which mainly takes place in the hygroscopic region, can be described using Fick's law, originally developed for unimpeded flow in air:

$$J_{diff} = -D_v \cdot \frac{M}{R \cdot T} \cdot \nabla p_v \tag{2.18}$$

where J_{diff} is the mass flux rate in kg/m^{2}s, D_v is the diffusion coefficient, R is the universal gas constant (8.314 J/mol K), M is the molar mass of water (0.018 kg/mol) and p_v is the partial vapour pressure. In porous materials, the flow is reduced compared to the flow possible in air due to the volume fraction of pores a and the pores' tortuosity α:

$$J_{diff} = -\alpha \cdot a \cdot D_v \cdot \frac{M}{R \cdot T} \cdot \nabla p_v \tag{2.19}$$

Most building codes introduce a resistance factor called μ to represent in a simplified way the tendency of building materials to avoid the water vapour to pass through, resulting in the following vapour diffusion equation:

$$J_{diff} = -\frac{D_v}{\mu} \cdot \frac{M}{R \cdot T} \cdot \nabla p_v \tag{2.20}$$

The coefficient D_v is here referred to as vapour permeability and is an intrinsic characteristic of the material in which the diffusion is taking place. If the medium is air, this value can be found using Schirmer's equation, where $D_{v,da}$ represents the binary diffusion coefficient between vapour and dry air:

$$D_{v,da} = \frac{2.26173}{P_a} \left(\frac{T}{273.15} \right)^{1.81} \tag{2.21}$$

If the medium is a building material, this value cannot be calculated but can be found empirically using the standard ASTM E96 as reference (ASTM E96/E96M-16, 2016).

Compared to diffusion, moisture transfer by convection may be much more significant in buildings. This mechanism can be described by the following equation:

$$J_{conv} = -\rho_v \frac{k \cdot k_{rg}}{\mu_g} \nabla p_g \tag{2.22}$$

where ρ_v is the vapour density expressed in kg/m^3, k is the permeability, k_{rg} is the vapour relative permeability, p_g is the vapor pressure and μ_g is the dynamic viscosity (Pa.s).

2.3.1.2 Liquid transfer

Porous materials are characterised by different regions of interest defined by the sorption isotherm. The hygroscopic regime is dominated by diffusion, while the capillary regime is where suction takes place. This region extends to the achievement of capillary water saturation, at which point suction processes are not feasible and temperature gradient or external pressure under suction drives the transfer. In this region, as seen, the state of equilibrium does not apply.

In the capillary region the liquid flow can be described using Darcy's law, extended to include the gravity force:

$$J_w = -\left(D_\phi \cdot \nabla RH + \frac{D_\phi \cdot \frac{\partial RH}{\partial w}}{\frac{\partial P_h}{\partial w}} \cdot \rho_w \cdot g \right) \tag{2.23}$$

where the D_ϕ liquid coefficient is expressed in kg/m s, g is the gravitational acceleration (9.81 m/s^2), RH is the relative humidity, w is the moisture content, and ρ_w is the density of water. P_h is the suction pressure, which can be described using a cylinder capillary model:

$$P_h = \frac{2 \cdot \sigma \cdot \cos(\theta)}{r} \tag{2.24}$$

where σ is the surface tension of water (72.75 10^{-3} N/m at 20°C), r is the capillary radius (m), and θ is the contact angle between the liquid meniscus and the pore wall. Considering formula 2.15, this term can also be expressed as a function of the relative humidity.

2.4 Combined heat, air and moisture transfer

2.4.1 Development of moisture transfer models

The development of a model that can account for all the moisture transfer mechanisms has been a topic of research since the mid-1900s and there is still no agreement on a single mathematical model. The foundation knowledge of this field was initially based on work conducted the field of soil mechanics (Hens, 2013), establishing a comparison between soil and porous materials. The first models were developed for unsaturated porous media with the effects of temperature (Luikov, 1964) and have been validated for transport in soil (Thomas, 1987) using moisture content as the driving potential. These first models assumed simplified boundary conditions, such as constant air pressure and moisture content discontinuity at the interface between two materials.

The first method for use in building applications, the so-called Glaser's method, was only developed in the 1980s. This is a simplified steady-state technique that considers vapour transfer for diffusion assuming low moisture content, constant material properties and constant boundary conditions (refer to Chapter 3). However, the numerous limitations of the Glaser's technique prompted researchers to develop a more refined mathematical model that could better describe the physical phenomenon. A subsequent model considered liquid flow at high moisture content levels by dividing moisture transport into its vapour and liquid components, thus coupling two different equations (Pedersen, 1992). Pedersen used capillary suction as driving potential, and this has also been considered in other models (Li, Rao, & Fazio, 2009; Rouchier, Woloszyn, Foray, & Roux, 2013). However this still fails to preserve the continuity at the interface of two different materials (Straube, Onysko, & Schumacher, 2002). In the search for a more suitable driving potential, other models have been developed. One model relies on the air humidity ratio (Budaiwi, El-Diasty, & Abdou, 1999) but only includes vapour transfer, neglecting the importance of capillarity and water diffusion. Other approaches have used relative humidity as driving potential (Steeman, Van Belleghem, Janssens, & De Paepe, 2009; Tariku, Kumaran, & Fazio, 2010). Of these, the model developed by Kunzel and Kiessel relies on the use of retention function from sorption isotherm and suction curves to perform and solve very complicated hygrothermal transfer issues. This model is among the most widely used in building engineering for 1D hygrothermal simulations, especially since the development of the software WUFI. A later approach elaborated the earlier models to include vapour and liquid transfer (Mendes, Philippi, & Lamberts, 2002) and solve the issue of discontinuity at the interface (Mendes & Philippi, 2005), but it proved too complicated for 3D applications (Dos Santos & Mendes, 2009) and, in common with all the above-mentioned models, ignored air transport. Häupl, Grunewald, Fechner, and Stopp (1997) included heat, air and moisture transfer in one model by using the moisture, air and energy conservation equations, with moisture content, air pressure and temperature, respectively, as driving forces. Another attempt to include the convective effect was made by Hagentoft et al. (2004), where air was assumed as a constant and permanent infiltration through 1D porous materials, while temperature and capillary suction were used as driving potentials for energy and mass transport. A subsequent development sought to address this shortcoming by including the ability to define variable air pressure and consider the convective process within multilayer envelopes (Dos Santos & Mendes, 2009). Numerous other models have been developed more recently to account for different boundary conditions, such as wind-driven rain and rising damp, or integrate refined hysteresis effects of moisture storage (Carmeliet, de Wit, & Janssen, 2005; Kwiatkowski, Woloszyn, & Roux, 2009; Lelievre, Colinart, & Glouannec, 2014; Ouméziane, 2011).

The existence of different models for comprehensively describing heat, air and moisture transfer makes it impossible to provide a definitive reference, as all these models have weaknesses and limitations. However, the basic logic is based on the same principles. All integrated HAM models combine heat, moisture and air transfer through different mathematical equations that try to solve the transport process simultaneously via conservation of energy for heat transport, mass for air and moisture transport, and momentum for air and vapour transport.

2.4.2 Hygrothermal assessment

The interactions between moisture and buildings greatly influences building performance. Condensation, durability, biodeterioration and indoor environmental quality are all determined by hygrothermal processes within the building and the building envelope. For example, sorption curves are essential to determine both the risk of condensation and the level of indoor environmental quality provided. However, the uptake in practice of the latest knowledge and hygrothermal calculation models has been slow at best.

Research on HAM models has been ongoing since the early 80s and has generated new knowledge and better understanding of how to describe moisture transfer. Nonetheless, the steady-state psychrometric approach is still the most widely used in practice. On one hand, advances in HAM modelling have helped to refine how moisture is accounted. On the other hand, this has led to the development of sophisticated models that are not feasible for application in everyday professional practice.

The following chapters analyse the different impacts of moisture on buildings, describe the available software and the level of knowledge needed to perform hygrothermal transient assessments, and investigate the policy framework in order to uncover the reasons behind the discrepancies between research and practice.

References

ASTM C1498-04a. (2010). Standard test method for hygroscopic sorption isotherms of building materials.

ASTM E96/E96M-16. (2016). Standard test methods for water vapor transmission of materials.

Budaiwi, I., El-Diasty, R., & Abdou, A. (1999). Modelling of moisture and thermal transient behaviour of multi-layer non-cavity walls. *Building and Environment*, *34*(5), 537–551. Available from https://doi.org/10.1016/S0360-1323(98)00041-9.

Carmeliet, J., de Wit, M., & Janssen, H. (2005). Hysteresis and moisture buffering of wood. In Proceedings of 7th Symposium on Building Physics in the Nordic Countries.

Dos Santos, G. H., & Mendes, N. (2009). Combined heat, air and moisture (HAM) transfer model for porous building materials. *Journal of Building Physics*, *32*(3), 203–220. Available from https://doi.org/10.1177/1744259108098340.

Hagentoft, C. E., Kalagasidis, A. S., Adl-Zarrabi, B., Roels, S., Carmeliet, J., Hens, H., ... Djebbar, R. (2004). Assessment method of numerical prediction models for combined heat, air and moisture transfer in building components: benchmarks for one-dimensional cases. *Journal of Thermal Envelope and Building Science*, *27*(4), 327–352. Available from https://doi.org/10.1177/1097196304042436.

Häupl, P., Grunewald, J., Fechner, H., & Stopp, H. (1997). Coupled heat air and moisture transfer in building structures. *International Journal of Heat and Mass Transfer*, *40*(7), 1633–1642. Available from https://doi.org/10.1016/S0017-9310(96)00245-1.

Hens, H. (2013). Moisture control in buildings. A Handbook of Sustainable Building Design and Engineering: An Integrated Approach to Energy, Health and Operational Performance.

Hens, H.S. (2017). Building physics-heat, air and moisture: fundamentals and engineering methods with examples and exercises.

Kwiatkowski, J., Woloszyn, M., & Roux, J. J. (2009). Modelling of hysteresis influence on mass transfer in building materials. *Building and Environment*, *44*(3), 633–642. Available from https://doi.org/10.1016/j.buildenv.2008.05.006.

Lelievre, D., Colinart, T., & Glouannec, P. (2014). Hygrothermal behavior of bio-based building materials including hysteresis effects: experimental and numerical analyses. *Energy and Buildings*, *84*, 617–627. Available from https://doi.org/10.1016/j.enbuild.2014.09.013.

Li, Q., Rao, J., & Fazio, P. (2009). Development of HAM tool for building envelope analysis. *Building and Environment*, *44*(5), 1065–1073. Available from https://doi.org/10.1016/j.buildenv.2008.07.017.

Luikov, A. V. (1964). Heat and mass transfer in capillary-porous bodies. *Advances in Heat Transfer*, *1*(C), 123–184. Available from https://doi.org/10.1016/S0065-2717(08)70098-4.

Mendes, N., & Philippi, P. C. (2005). A method for predicting heat and moisture transfer through multilayered walls based on temperature and moisture content gradients. *International Journal of Heat and Mass Transfer*, *48*(1), 37–51. Available from https://doi.org/10.1016/j.ijheatmasstransfer.2004.08.011.

Mendes, N., Philippi, P. C., & Lamberts, R. (2002). A new mathematical method to solve highly coupled equations of heat and mass transfer in porous media. *International Journal of Heat and Mass Transfer*, 509–518. Available from https://doi.org/10.1016/s0017-9310(01)00172-7.

Ouméziane, Y.A. (2011). Hygrothermal behaviour of a hemp concrete wall: influence of sorption isotherm modelling.

Pedersen, C. R. (1992). Prediction of moisture transfer in building constructions. *Building and Environment*, *27*(3), 387–397. Available from https://doi.org/10.1016/0360-1323(92)90038-Q.

Rouchier, S., Woloszyn, M., Foray, G., & Roux, J. J. (2013). Influence of concrete fracture on the rain infiltration and thermal performance of building facades. *International Journal of Heat and Mass Transfer*, *61*(1), 340–352. Available from https://doi.org/10.1016/j.ijheatmasstransfer.2013.02.013.

Steeman, H. J., Van Belleghem, M., Janssens, A., & De Paepe, M. (2009). Coupled simulation of heat and moisture transport in air and porous materials for the assessment of moisture related damage. *Building and Environment*, *44*(10), 2176–2184. Available from https://doi.org/10.1016/j.buildenv.2009.03.016.

Straube, J., Onysko, D., & Schumacher, C. (2002). Methodology and design of field experiments for monitoring the hygrothermal performance of wood frame enclosures. *Journal of Thermal Envelope and Building Science*, *26*(2), 123–151. Available from https://doi.org/10.1177/0075424202026002098.

Tariku, F., Kumaran, K., & Fazio, P. (2010). Transient model for coupled heat, air and moisture transfer through multilayered porous media. *International Journal of Heat and Mass Transfer*, *53*(15–16), 3035–3044. Available from https://doi.org/10.1016/j.ijheatmasstransfer.2010.03.024.

Thomas, H. R. (1987). Nonlinear analysis of heat and moisture transer in unsaturated soil. *Journal of Engineering Mechanics*, *113*(8), 1163–1180. Available from https://doi.org/10.1061/(ASCE)0733-9399(1987)113:8(1163).

Durability, condensation assessment and prevention

3

Durability, or the capacity of a building to maintain its functionality over time, has been a relevant topic in the construction sector from its early development. However, it has grown in importance in recent decades due to rising expectations in relation to building performance and the development of new construction techniques. Traditional construction techniques and technologies relied on common design practices where heavyweight envelopes proved suitable for the specific contexts in which they were employed. The introduction of lightweight, multilayered and multifunctional envelopes challenged this approach, prompting refection on the interconnections between design, materials and durability. Multilayered envelopes rely on the concept of combining different materials to respond more effectively to varied requirements rather than having all functional responses dependent on a single layer. Massive traditional monolayer envelopes have the capacity to resist deterioration regardless of the presence of small defects in the construction, lightweight envelopes, instead, require increased resilience and redundancy to prevent construction failures. When performance requirements are assigned to different functional layers, it is of utmost importance to design each layer carefully to optimise the performance of the whole envelope as a holistic system.

The majority of durability failures are found to be moisture related, due to bad construction and design practices, poor maintenance or inappropriate operation and usage of the building. Since condensation is the main cause of such problems, it is essential to understand its effects on the overall building envelope, including corrosion of metallic elements, and the consequent risk to structural integrity and reduction of thermal performance. The latter is due to water reducing the thermal resistance of hygroscopic insulation materials, leading to aesthetic damage, rot and significant health hazards due to the growth of mould, bacteria and mites. All these effects produce increased risks, as well as increased costs for repair, maintenance, insurance and litigation related to moisture damage claims.

The risk of condensation affecting a building envelope is influenced by the material selection, the overall composition of layers and the climatic pressure on both sides. While little can be done to influence the exterior climate, the other two components can be controlled during the design and operating phases, where there are opportunities to increase hygrothermal performance. For this, designers, constructors, occupants and policymakers have a shared responsibility.

A moisture-aware design process is fundamental to reducing the risk of condensation, improving the building's hygrothermal performance and, consequently, ensuring durability. Design strategies, materials selection, and envelope design need to be carefully tailored to the building's current and future needs. This approach

Moisture and Buildings. DOI: https://doi.org/10.1016/B978-0-12-821097-0.00006-0

should also include strategies to increase the building's resilience and capacity to adapt to changing conditions and provide a certain level of redundancy for maintaining function during unexpected failures.

However, design alone is not sufficient to minimise condensation risks, as the construction quality significantly influences the building's hygrothermal performance, especially in relation to the continuity and integrity of thermal, water and moisture barriers. Further, once the building is inhabited, occupants have the same level of responsibility to operate the building correctly. Indoor moisture generation and underventilation of spaces are among the major causes of condensation and moisture-related issues, especially in new buildings, with consequent destructive effects on the building envelope and internal finishes. Hence, understanding the impact of these factors on the building's hygrothermal performance is not optional, and occupants should be educated about their role in reducing the condensation risk. A second level of resilience must be provided through a strategically planned preservation agenda, as recurrent interventions in the form of regular maintenance are significantly more effective than ad hoc repairs to extend the building's service life.

Policymakers also have a responsibility along with those involved in the building's design, maintenance and occupation. Durability is heavily influenced by regulations and requirements that are often developed to respond to other needs but which fail to acknowledge their impact on the overall hygrothermal performance of buildings. For example, the increased airtightness requirements associated with improving energy efficiency and addressing fireproofing contingencies, when not coupled with adequate ventilation and dehumidification strategies, contribute to an increase in indoor moisture content. This, in turn, increases the hygrothermal pressure on the envelope, which is often unable to respond appropriately because it has been designed with different assumptions and to withstand different hygrothermal stresses.

The application of physical principles of hygrothermal processes to the building environment can help to describe and predict the hygrothermal behaviour of the building envelope in relation to its climatic context, both interior and exterior. By understanding the interactions between moisture and building envelopes, as well as the role of design in determining and controlling their implications, it is possible to reduce the occurrence of moisture-related issues and provide better performance and more durable buildings. This chapter investigates the issues around durability, explores moisture-related durability risks, and explains the various mathematical models that designers can use to predict or analyse the hygrothermal behaviour of building elements.

3.1 Durability

Durability can be defined as the capacity of a building material, component or structure to perform its function over an intended service period under the influence of degrading mechanisms. As such, durability is not a specific property of a material or component but, rather, the outcome of the interactions among several factors, such as climate, construction quality, design practices and usage. From a holistic perspective, buildings without defects and durability failures require less frequent repair and maintenance, thus reducing the overall environmental load associated with the embodied energy of the new materials, the construction waste generated and the reparation process itself (Maia, Ramos, & Veiga, 2019). This additional embodied energy is usually accounted for in life-cycle assessment via the recurring energy parameter (ISO 14040, 1997). Integrating durability as a design parameter may help to reduce this component and ensure more sustainable buildings.

Designing for durability does not mean designing buildings that will last forever but, rather, understanding the specific requirements of each building element and calibrating its durability performance in relation to aesthetics, safety and health. This process involves defining the minimum design life needed for the building, its components and materials to retain their function. The value of the design life changes according to the building code used as reference, but three general categories can be identified (Australian Building Code Board, 2015):

- Short lifespan: temporary structure or replaceable structural parts are usually designed for a lifespan that varies between 1 and 25 years.
- Normal lifespan: building structure and common construction are usually designed for a lifespan of 50 years.
- Long lifespan: key infrastructures, such as train stations and airports, bridges or civic buildings, have a design life of 100 years.

Durability involves the whole building in all its subsystems, including structures, envelope, mechanical systems and finishes. In a holistic approach to building design, one element cannot be separated from the others, as all strongly influence the performance of one another. A second layer of complexity is added by the fact that durability concerns the building over its entire life cycle: from concept design to construction, from operation to long-term maintenance. A robust, consistent and up-to-date preservation plan is as important as a design that considers durability performance as part of its key requirements. Designing for durability means accounting for the factors that influence the performance of a building or component and planning for regular interventions aimed at retaining performance over time.

3.1.1 *Main actors involved in durability*

As durability is a concept that underpins the whole life cycle of a building, it is a concern for all those with an active role in building design, construction and operation (Soronis, 1992), namely:

- *Occupants* have a considerable impact on the way buildings are used and operated, which ultimately influences some of the degrading parameters and mechanisms that combine to determine the service life. In Fig. 3.1, these actors influence the inclination of the curve.

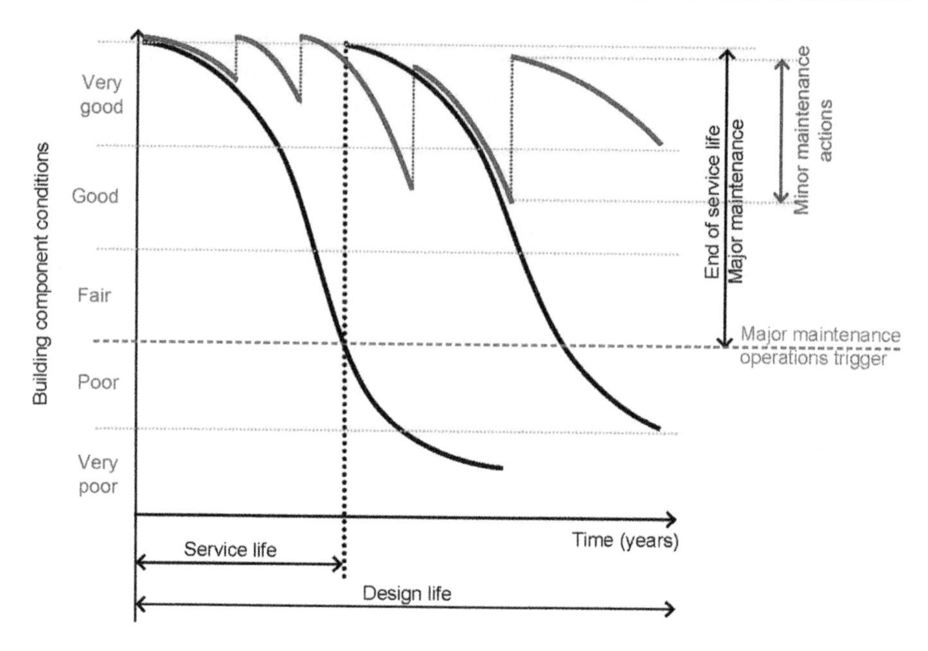

Figure 3.1 Curve of performance as a function of time: without maintenance, with extraordinary repair and with regular maintenance. The service life of a building (components or material) is a function of the durability programme established during the early stage of its life.

- *Designers* (architects and engineers) have control over the conceptual and design phase of buildings, during which they usually make assumptions about the occupants' behaviour and the future use of the building. Architectural choices, such as materiality and geometry, and engineering choices, such as heating, ventilation and air conditioning (HVAC) and plumbing design, must be underpinned by an overarching durability strategy to provide the best starting point. In Fig. 3.1, these actors determine the inclination of the curve and the overall behaviour of the building.
- *Contractors*, including constructors and suppliers. Construction defects can significantly impact the durability performance of buildings, so it is essential to ensure that the building is constructed according to the design's assumptions. Further, particular care must be taken during this phase in relation to the specific products that are installed, since they have a direct influence on the building's performance. In Fig. 3.1, contractors influence the inclination of the curve.
- *Property managers* take over the building after construction and engage with maintenance personnel and occupants to ensure that the preservation plan is adhered to. The importance of these actors is highlighted in Fig. 3.1, where they are responsible for the overall trends over time.

All these actors have great potential to influence the final durability performance of buildings. Property managers and occupants can ensure the rigorous application of the preservation plan, thereby significantly extending service life defined as 'the period of time after installation a product is expected to meet all its minimum functional requirements when routinely maintained but without major repair being necessary'. Although designers and contractors cannot directly affect the operational phase, they have the power to ensure durability is accounted for during the early design and construction phase by using an informed approach to selection, integration, construction and design of building components, materials and equipment. It is worth noting that new materials and construction methods require additional care in the context of durability: it is essential to understand the new systems, their performance and their implications for trades and quality control (Soronis, 1992), since these may directly or indirectly affect all actors.

3.1.2 Degradation factors and mechanisms in durability

The three main systems that determine the overall durability performance of a building are as follows: its envelope and building components system, including materials and finishes; the structure; and the mechanical equipment, including plumbing installations. Each of these is subject to different degrading mechanisms from the factors that can affect the response to durability issues. These include the following:

- Time: decay, corrosion, sustained stress, operation.
- Natural hazards: hurricanes, flooding, bushfires, earthquakes.
- Biological hazards: insects, mould, bacteria.
- Chemical agents: pollutants.
- Natural elements: groundwater, soil type, salinity (exposure to airborne salt).
- Climatic factors: rainfall, heat, solar radiation [especially ultraviolet (UV)], wind, moisture.

The intensity, concentration and frequency of these factors influence the severity of the durability issues.

Time is a major determining factor in overall durability performance. Some materials and systems are more subject to loss of performance over time and therefore have a shorter lifespan. Understanding the real lifespan of these systems and designing accordingly is a major challenge in designing for durability. The added value of a durable building does not lie in selecting systems with a longer lifespan but in understanding the scenarios in which a shorter period is acceptable and how to design for easy replacement or repair of this system.

In contrast to time, natural hazards are not an expected occurrence during the lifetime of a building. However, it is important to provide a certain degree of resilience and ensure that buildings can withstand exceptional events. It is worth noting that climate change and human activities are increasing the frequency and intensity of some natural disasters, such as bushfires and tornados (Vardoulakis, Marks, & Abramson, 2020). In this context, it is of utmost importance to account for the possibility of a natural disaster. Usually, building codes address this issue by establishing minimum levels of disaster mitigation for public safety and general welfare.

Biological hazards include mould (refer to Chapter 4: Health and mould growth), bacteria and insects. Some insects, such as termites, carpenter ants and carpenter bees, have a significant impact on building materials, especially biological or bio-based products. While a degree of protection can be provided by chemical soil treatments and insect shields, the probability of an insect infestation should be carefully considered during the design and construction phase (Newport Partners & ARES Consulting, 2015).

Other factors that can influence the early degradation of building materials are chemical agents and atmospheric pollutants. Chemical reactions between volatile organic compounds and nitrous oxides under the effect of sunlight can create ozone. Ozone is beneficial in the stratosphere, thanks to its capacity to block UV radiation, but it may be a hazard when formed closer to the terrestrial crust. Indeed, it not only poses a health risk for humans, especially in regard to respiratory issues, but also has an effect on many synthetic materials, including rubber, polyester, cotton textiles and certain paints. Similarly, sulphur dioxide in the atmosphere is the primary cause of acid rain, which can corrode materials such as limestone panels, copper and various roofing materials, contributing to a reduction in the lifespan of these building components.

Among the natural factors, ground and surface water are the most widespread source of early degradation of building components. A high water table and poor drainage can lead to foundation cracking, rising damp and other moisture-related damaging mechanisms. In particular, the loss of draining surfaces in urban environments in favour of asphalt or other impermeable materials can easily lead to flooding and direct the excess water towards buildings, causing severe damage to the envelope and increasing the risk of moisture-related issues, such as oxidation, corrosion and mould growth. A long-term vision for city development and building design accounts for this potential by balancing the impermeable surfaces with parks and green areas that can drain rainwater during extreme events. Climate is another important factor in determining the service life of building systems. For example, solar radiation greatly impacts the durability performance of materials. The sun radiates energy that reaches the earth along a wavelength spectrum, of which visible light is only a small portion. UV and infrared (IR) light are among the nonvisible wavelengths that have major effects on buildings, causing either chemical or physical reactions. IR determines materials' expansion and contraction, which can influence their long-term performance, especially in complex systems where different components have different thermal expansions. On the other hand, UV is responsible for a broader range of effects that cause materials to fade, became brittle and crack (Ashton & Sereda, 1981).

Among all the factors that represent a risk to durability performance, moisture is regarded as the trickiest. Almost 70% of construction defect claims are associated with moisture (Oppenheim & Brennan, 2008), with estimates of the economic impact of repairs at around 10 million US dollars (Newport Partners & ARES Consulting, 2015). Recent estimates suggest that this figure may be even higher (Dewsbury, Law, & Henderson, 2016; Law and Dewsbury, 2018) when more general moisture-related repairs are taken into account. The *3Ds rule – deflect, drain and dry* – usually provides good guidance on the design principles that can be used to manage moisture in buildings. These include the use of rainscreens to protect the building envelope from rain and wetting, and the use of water-resistant barriers to create a waterproof plane that can protect the insulation layers, thus enhancing drying capacity. While this rule may appear simple, its application is anything but easy.

Ultimately, durability performance depends on the combination of all these different factors, which interact with one another over the whole life cycle of a building (Lacasse, 2003). These interactions are difficult to predict. Knowledge of the various degradation mechanisms and impacts on building systems can inform a design approach aimed at extending the building's service life. Among all the degradation factors and mechanisms, moisture management is to date least understood and considered during the design stage.

3.1.3 Prediction of service life

Designing for durability means taking into account all the possible interactions between the various factors that influence durability performance to establish both design strategies, aimed at minimising the risks, and a coherent preservation plan. This process relies on the service life of each component, material or system that is part of the building and influences its final overall durability performance. Hence, the prediction of service life is essential to guarantee durability and to conceptualise deterioration (Masters & Brandt, 1989). This can be done by adopting one or more of the following methods (Lewry & Crewdson, 1994):

- Comparison with similar cases, where the similarity must be established in relation to construction techniques, occupation and climatic context. This method is largely based on previous and personal experience and, as such, can be employed at the beginning of the design phase as a general principle. However, it is highly subjective and not project-specific and is therefore the least accurate and reliable method.
- Assessment of the short-term deterioration rate during the first period of use or exposure and projecting the performance over a longer period. This method measures the real use and exposure context, but its reliability over the long term is undermined by the assumption that the conditions of use and exposure do not change over time. However, this method can be used in association with others to refine the prediction, as well as for continuous assessment during the operational life to update the information and strategies included in the preservation plan.
- Interpolation of the results obtained through accelerated tests. These experimental procedures are quite complex and the definition of the assumptions and boundary conditions is a delicate operation that determines the reliability of the results. Hence, experienced testers are required to ensure the tests are correctly applied.

None of these methods is precise or accurate (Soronis, 1992), but they represent a reliable starting point for the development of strategies aimed at improving building durability. The accuracy of the results can be increased when all three methods are employed at different stages of the building life cycle to inform the design and the preservation plan and to update the repair and maintenance schedule. Designing for durability does not mean designing a building, but rather developing a design solution and a coherent preservation plan.

A systematic approach to service life prediction may help to increase the reliability of the results (Masters & Brandt, 1989). A few methodologies address this issue by either introducing a model to identify the relevant parameters to be considered in the prediction (Lucchini, 1990), or standardising the whole prediction process through a series of iterative steps that include defining the scope of the analysis and identifying the key deterioration parameters, testing procedures and analysis of the results (Lacasse, 2003).

3.2 Moisture and durability

Moisture is broadly acknowledged as a key element in durability as it is involved in most of the deterioration, biodeterioration and corrosion processes, either as a direct agent or as a medium of the reaction (Ashton & Sereda, 1981). In some of these processes, even small traces of moisture, such as thin surface films, are enough to initiate the deterioration process; prolonged exposure, cyclical conditions and interactions with other factors can speed up the process, leading to failure.

Moisture, as a degrading mechanism, involves the three states of water: vapour, liquid and solid. Freezing and thawing, rain, moisture-laden air and vapour equally influence the durability performance of building systems. Some of these factors can be controlled more easily and intuitively than others.

Rainwater management, for example, can be addressed with appropriate design strategies and detailed information about the geometry, size, depth and materials of the roofing system. As long as flashing, water barriers and sealing systems are carefully executed during the construction phase following high-quality standards, and water leakages are prevented, degradation due to rainwater can be effectively controlled. Reducing the availability of liquid in the envelope can also improve the building's response to freezing and thawing. Moisture-laden air is usually managed through the use of air-barriers that wrap the building envelope and seal the indoor environment. Airtightness is an increasingly important requirement in building codes around the world for improving the overall energy efficiency of the built environment. In fact, the introduction of different barriers in the building envelope is emerging as the main strategy to control moisture and water movement. These membranes are designed to allow certain exchanges between the indoors and the outdoors, while preventing others from occurring, based on their permeability.

Intuitively, these membranes should be placed to protect the whole envelope from the external environment, that is, behind the cladding or rainscreen. However, vapour barriers are not the right solution for all climates, and different degrees of permeability are required in different contexts (Derome, 1999). Vapour management cannot be addressed by a 'ticking-the-box' approach or a list of deemed-to-satisfy criteria, as the issue is embedded in the complexity of the building materials' and systems' hygroscopic behaviour. Failure to specify the right level of vapour permeability of the envelope, the placement of the vapour barrier and selection of material can easily lead to condensation inside the building envelope, ultimately influencing durability performance. Hence, it is of utmost importance to understand the hygrothermal behaviour of the building envelope and the condensation phenomenon to ensure durability and prevent the occurrence of major moisture-related failures.

3.2.1 *Effects of condensation*

Generally, condensation can occur on the surface or inside the building envelope. The former is caused by warm and moist air that comes into contact with a cooler surface, usually at or below the air dew point. This is referred to as surface condensation and is typically observed on windows during winter, when the indoor area is heated but the surface temperature of the glass is much lower. The second case is represented by so-called interstitial condensation, which occurs within the building envelope. Interstitial condensation is caused by vapour diffusing within the building materials from a warm and humid environment to a cooler and drier one. If the temperature drops significantly across the envelope and the dew point is reached, vapour forms condensation within the materials.

Interstitial condensation is more difficult to detect, control and manage due to the fact that it is visible to occupants only when the water accumulation or its effects have reached critical thresholds that allow it to be seen outside the envelope. The effects of condensation are destructive (Dewsbury, Law, & Henderson, 2016) and highly important, since they can be direct or indirect, as well as visible or invisible. These effects include early deterioration of building materials (Allsopp, Seal, & Gaylarde, 2004), support of biological growth (Ortega-Calvo, Hernandez-Marine, & Saiz-Jimenez, 1991), reduced overall thermal performance (Cai, Zhang, & Cremaschi, 2017; Ogniewicz & Tien, 1981) and, ultimately, reduced service life.

Water molecules are strongly polarised, making water a perfect solvent and chemical catalyst and leading to structural and chemical issues. For example, binders in particle boards, plywood and oriented strand board are subject to hydrolysis, a chemical reaction induced by water that captures the molecular bonds of other substances. The direct effects of this reaction are losses in structural strength and stiffness, which can lead to cracking and reduced ability to carry loads. Certain building materials in marine environments can quickly degrade if condensation hydrates the salt deposited on their surface, corroding metallic elements, such as concrete rebars, and degrading stone (Stefanoni, Angst Ueli, & Elsener, 2018). Water is also fundamental to support mould, mildew, fungi and bacteria growth (refer to Chapter 4: Health and mould growth), which not only has aesthetic implications but is also associated with serious health hazards, especially respiratory illness, and structural risks. Fungi can digest the building materials and completely degrade the building envelope within a very short period of time, given the right combination of water, nutrients and temperature. Similarly, moss and algae can take root inside mortars and joints, degrading the structural integrity of the building envelope, while the organic acids that are produced can trigger metal corrosion (Ortega-Calvo et al., 1991). Furthermore, condensation can significantly reduce the thermal performance of the building envelope. Hygroscopic thermal insulations, when wet, are subjected to increased latent heat exchanges that reduce the thermal resistance value (Cai et al., 2017; Ogniewicz & Tien, 1981).

3.3 Condensation in buildings

As previously explained, condensation is defined as the phenomenon that occurs whenever a surface is at a temperature lower than the dew point of the air that surrounds it, or whenever vapour, moving across the envelope, comes into contact with a material at a lower temperature than its dew point. Condensation can be the result of either vapour diffusion or moisture-laden air movements and can be caused by migration of air and vapour from an environment with higher vapour pressure to another with lower vapour pressure. These two environments are usually the indoors and outdoors, but the phenomenon can occur in any situation where a component divides two spaces characterised by different climates, such as internal partitions between bedrooms and bathrooms, or dividing walls between two different residential units. The vapour flux direction can be determined from the assessment of the vapour pressure conditions in the two environments separated by the envelope. For many years, condensation was regarded as a phenomenon that only relates to cold climates (Australian Building Code Board, 2014a) characterised by cool outdoor environments and heated indoors. The heating system, coupled with the moisture generated by the building occupants and low ventilation rates, produces higher vapour pressure indoors, hence vapour flows outward. However, the widespread deployment of air conditioning systems on a large scale, especially when not coupled with a dehumidification system, has resulted in the increasing occurrence of the opposite problem, that is condensation occurring in warm climates, further aggravated by the lack of regulations around minimum set point temperature for HVAC systems in the residential sector during summer. Air conditioning systems can reduce both temperature and relative humidity compared to the exterior environment, creating a significant difference in vapour pressure. In this case the vapour flux is directed inward and, consequently, the location of the critical surface in the envelope differs from that in cold climates. Temperate climates may be subject to both conditions depending on the season.

Determining the direction of the flux is of utmost importance for understanding where and what type of vapour barrier, if any, can be installed. Considering interstitial condensation as the result of moisture-laden air moving in a certain direction and forming dew when suddenly exposed to lower temperatures, it is clear that this drop in surface temperature can occur in proximity to the thermal control layer or insulation. In cold climates, or heating seasons, condensation is most likely to occur on the interior side of the envelope. In warm climates, or cooling seasons, condensation is most likely to occur on the exterior, when warm outdoor air comes into contact with surfaces cooled by the air conditioning system.

In all instances, the vapour barrier should be placed on the warmer side of the insulation layer to prevent vapour from reaching the colder materials. However, this is not always effective in mitigating condensation risk, as seasonal variations in the moisture pressure flux direction can easily lead to condensation on the opposite surface (Lstiburek, 2002). Hence, an early hygrothermal assessment is essential for designing buildings and envelopes that minimise the risk of condensation. The first step in this direction is to have a good understanding of the causes of condensation.

3.3.1 Parameters that influence the risk

The parameters that influence the risk of condensation can be categorised into design, construction, occupants and environmental. Of these, the environmental category is usually regarded as the most important, and external climatic conditions are regarded as the major driver in condensation risk (Hens, 2017). However, the design and construction phases, as well as the maintenance and occupation patterns, are just as important and offer a higher degree of control and manageability.

3.3.1.1 Moisture sources

The main moisture sources are the indoor and outdoor moisture content, and the moisture trapped within the envelope during construction.

3.3.1.1.1 Outdoor moisture sources

Outdoor sources of moisture are all those events and parameters related to the climatic conditions. Outdoor relative humidity, rain, fog, mist and soil-generated moisture are all part of this category. These moisture sources cannot be controlled directly by the various actors who determine the building's hygrothermal performance over its life cycle, but they can be predicted with reasonable accuracy through statistical analysis of the climatic context. Nonetheless, the quality of the construction and the execution of the preservation plan are vital to preventing uncontrolled water ingress due to failures and defects over time. This source of additional moisture includes leaking pipes, water penetration through gaps and cracks, and rising damp from the terrain (Hens, 2013).

Accidental rain penetration is usually managed with the first two Ds of the 3Ds rule (deflect and drain) but, when small and frequent ingresses are present – which usually results from construction defects – the consequences for the envelope may be critical. Similarly, leaks of moist air can lead to interstitial condensation or increased water availability on the surfaces, which may eventually initiate mould growth and biodeterioration. Further, if timber or other moisture-sensitive materials are present, these accidental moisture sources may also cause rot, corrosion, rust and general degradation. Below-grade or close to the terrain building components may also absorb additional moisture through rising damp and groundwater intrusions. The latter can be caused by a high water table or simply by rainwater soaking the ground. Particular attention must be given to crawl spaces, which have been found to be particularly prone to high humidity levels and mould growth (Keskikuru et al., 2018; Matilainen & Kurnitski, 2003). The cool and humid air trapped between the bottom of the floor and the ground infiltrates into the building through cracks, increasing the indoor humidity.

Another accidental moisture source is water from damaged piping systems. Plumbing system failures can lead to a significant amount of additional moisture. Even if it is impossible to manage large-scale failures at the design stage, the envelope should provide resilience by allowing for drying or deflection.

It should be noted that climate change is altering outdoor conditions, with rising temperatures and increased weather events, such as typhoons and heavy rains. These changing conditions need to be reflected in the assumptions underpinning the tools used to determine the role of outdoor moisture sources in the risk of condensation (Hao et al., 2020; Roberts, 2008).

3.3.1.1.2 Construction moisture

Moisture trapped during construction is only likely to affect the hygrothermal behaviour of a building in the first 2 to 3 years, by which time, under standard conditions, the envelope should have dried out (Hens, 2013). This initial moisture content is partly contained in new materials, such as concrete or timber, and partly due to uptake during the construction process, when materials are stored outdoors on the construction site. Although materials are usually covered and protected from rain, they can still come into contact with the terrain and be exposed to ground-source moisture. In addition, if water deposited on the slabs and uptake from the structure during construction has not dried completely before the façade is erected, the drying process will increase the indoor humidity and generate an additional source of moisture.

Furthermore, all wet trades, such as plastering and screeding, can release additional moisture into the air. During the curing and drying process, the moisture contained in this material evaporates into the air and, if the building is not designed to accommodate the extra amount of moisture, the vapour pressure within the envelope can increase significantly and lead to condensation. For example, it is estimated that each cubic metre of concrete releases over 90 L of water over the first 2 years after pouring and that, for new houses, the total moisture released in the first year averages 10 L/day, decreasing to 5 L/day in the second year (Hens, 2013). Although this process is time-limited, the consequences can include mould growth and rapid deterioration if moisture-sensitive materials are used.

3.3.1.1.3 Indoor moisture sources

Indoor moisture generation is among the most critical parameters in determining the hygrothermal performance of a building.

Humans produce moisture by breathing and transpiration from the skin; the rate of production depends on their emotional state, activity level and metabolism (Johansson, Pallin, & Shahriari, 2010). Humans also influence air humidity indirectly through the various actions they perform indoors which generate moisture, such as cooking, bathing and drying the laundry.

However, none of these sources can be considered stable in the long term, nor in the short period of daily variations. For example, bedrooms are a constant source of

Table 3.1 Average moisture generation rates for standard indoor activities.

Activity	Generation rate
Cooking	3 kg/day or 170–780 g/event
Washing laundry	0.5 kg/load
Drying laundry	5 kg/load or 450–2300 g/event
Dishwashing	Washing machine: 1 kg/day or 200–400 g/event Hand washing: 100–600 g/event
Personal hygiene	2.64–3 kg/h or 200–400 g/event
Breathing and transpiration	Sleeping: 72 ± 50 g/h Light activity: 30–120 g/h Medium activity: 79–200 g/h Intense activity: 200–300 g/h
Gas heating	0.5–1 kg/h
House cleaning	10–100 g/m^2 floor
Watering plants	0.07–0.5 g/h

moisture production during the night when people are sleeping, but the rate decreases to almost zero during daytime when they are not in use (Ilomets, Kalamees, & Vinha, 2018). In other rooms, the moisture generation may be driven by the water-related activities performed; in this case the moisture production rate is higher yet often limited in time (Zemitis, Anatolijs, & Frolova, 2016). The bathroom is a good example of water activity-driven moisture release that is highly variable over time. In this area, moisture is generated in short, intense periods of activity, such as showering, followed by a slow and prolonged phase during which water on the floor or condensed on surfaces evaporates, thus continuing to increase air humidity even after the initial source is turned off (Kalamees, Vinha, & Kurnitski, 2006). Another moisture-intensive activity is laundering. The practice of hanging wet clothes to dry indoors can release an amount of vapour between 1.22 (Zemitis et al., 2016) and 1.5 kg/day (BS, 2002), depending on the number of persons living in the apartment.

Table 3.1 shows the average amount of vapour generated by standard indoor activities (Yik, Sat, & Niu, 2004; Kalamees, Vinha, & Kurnitski, 2006; Johansson, Pallin, & Shahriari, 2010; Pallin et al., 2011; Ilomets, Kalamees, & Vinha, 2018).

When all these activities are taken together, the importance of the ventilation system in preventing the accumulation of indoor moisture is clear. In underventilated buildings, indoor moisture generation sources are the main drivers of condensation-related issues.

The events of 2020 revealed an additional layer of complexity in indoor moisture management. For example, Australia experienced an unprecedented bushfire season that lasted from June 2019 to March 2020, with incredibly high values of outdoor pollution. The public advice in these situations is usually based on the 'stay indoors' policy, which results in low air change rates as occupants attempt to protect the indoor environment form the intrusion of outdoor pollutants. Indirectly, this contributes to an increase in the indoor relative humidity and places additional hygric pressure on the envelope. Similarly, the global pandemic forced the world's population to remain inside their homes for a prolonged period of time, performing activities that are not usually performed indoors, such as working out, or simply staying inside more than usual. Although this additional moisture load is limited in time compared to the whole life cycle of a building, the risk of cyclical periods of lockdown (anticipated at the time of writing), or even the increasingly common practice of working from home, call for a review of the assumptions about indoor moisture generation that usually inform the design, assessment and management of moisture risks.

3.3.1.2 Design, construction and operation practices

Construction and design practices impact the building's hygrothermal behaviour by determining the overall response to the moisture pressure generated by humidity levels on the two sides of the envelope. However, these practices are continuously evolving to reflect changing comfort requirements and technological advances, thereby changing the envelope's hygrothermal performance. It is of utmost importance to acknowledge this evolution and consider materials and systems that influence the condensation risk.

3.3.1.2.1 Construction quality and process

In the architectural process, material and technological choices are driven by multiple factors, including the architectural intent, sustainability, environmental benefits, product availability, local construction practices, code requirements and cost. These different needs and provisions seldom work in the same direction but, rather, are often in competition with one another. This results in a process of negotiation and multioptimisation in which, if a performance is not strictly regulated by a building code or standard, it is simply overlooked. This is often the case for hygrothermal performance, which enters the picture only at a later stage, if at all, and only as a control check rather than a design driver. This issue is mainly due to the lack of a strong and detailed standard (refer to Chapter 7: Building codes and standards), lack of knowledge about condensation risk prevention, and the fact that moisture-related issues may manifest after a medium- to long-period of operation, thus reducing the perceived importance of the design strategies at the time of its occurrence. Further, the complexity of hygrothermal processes makes it difficult to establish clear and common guidelines for designers to bridge the existing normative gap (Mjornell, Arfvidsson, & Sikander, 2012).

The issue cascades directly into the construction phase. In the residential sector the design documentation often fails to specify the sarking system, the technological details and the required characteristics of the products (Dewsbury, Law, &

Henderson, 2016), leaving the choice to the constructors. Similarly, the documentation often fails to provide appropriately detailed descriptions of thermal insulation, weather and vapour membranes, which are a significant source of uncertainty in condensation prevention. Failure to clearly assign responsibilities in the design and construction process creates a gap in the system, leading to inadequate hygrothermal design solutions.

This gap opens up the possibility for the implementation of a value management process that may involve the use of cheap materials, such as chipboards or particle boards. There is nothing wrong with these materials, but the conditions in which they are used may introduce a significant source of hygrothermal risk due to their high susceptibility to moisture (WHO, 2009). It is essential to understand the hygroscopic behaviour of building products to specify the appropriate location and conditions of use to reduce the risk of serious moisture-related hazards (Aisner, Schimpff, Bennett, Young, & Wiernik, 1976).

One of the major causes of moisture-related durability issues in buildings is the presence of thermal bridges (Brambilla & Sangiorgio, 2020). A thermal bridge is formed when there is a sudden change of thermal resistance of the building envelope, usually produced by geometrical or technological factors that lead to a full or partial discontinuation of the thermal protective layer (Ilomets & Kalamees, 2016). The difference in thermal resistance will alter the heat exchange between two adjacent zones, which ultimately leads to localised microclimates and hygrothermal conditions on the material surface and within the envelope (Hanafi et al., 2018)(Santos et al., 2009) Only a few centimetres of additional external insulation in the presence of a thermal bridge can significantly reduce the risks (Fantucci, Isaia, Serra, & Dutto, 2017). However, this is not possible in all cases, as some retrofit interventions do not allow any modification of the external façade. For this reason, extensive hygrothermal analysis of thermal bridges should be undertaken during the design stage on a case-by-case basis to assess and minimise the risk of moisture-related durability issues (Bliuc, Lepadatu, Iacob, Judele, & Bucur, 2017).

While the design and construction processes can already introduce a higher risk of condensation and moisture-related issues, poor craftsmanship can magnify the risk. For example, mortar bridges in brick walls have been found to have a significant impact on the overall durability performance of the envelope (Calle, Coupillie, Janssens, & Van Den Bossche, 2020), as can the widespread practice of cutting the barriers for electric wiring and piping. The cumulative impact of all these factors can result in destructive condensation issues that could have been prevented and managed if they had been fully understood.

Table 3.2 shows some common moisture-related risks, and the potential causes behind those risks divided by design, construction and operation phases.

3.3.1.2.2 New design and construction trends

New energy efficiency standards rely on airtightness and increased levels of insulation as major strategies to reduce heat dispersion or gain through the envelope by minimising uncontrolled air and thermal exchanges between the interior and the exterior environments. However, higher levels of both insulation and airtightness minimise the opportunity for evaporation to dry the building fabric (Newport Partners & ARES Consulting,

Table 3.2 Most common causes of moisture-related risk divided by design, construction and operation phase.

Cause of condensation	Design	Construction	Operation
Air infiltration and leakages	Building not designed to operate in positive (or negative depending on the climate) pressure	Air barrier installation, cracks and defects – poor craftsmanship Cut and holes in barriers to allow for electrical wiring or plumbing	Poor maintenance of the envelope and system
Accumulation of moisture in the envelope	Envelope design not suitable; no drying opportunities Poor vapour barrier design Inappropriate vapour barrier permeability Poor thermal barrier design	Poor vapour barrier installation Construction defects in the flashing and water management systems Inappropriate vapour barrier permeability Thermal bridges and poor installation of the insulation layer	Poor management of the indoor environment Poor indoor ventilation, moisture-intensive activities Poor maintenance of the envelope and system
Accumulation of moisture indoor	HVAC design not accounted for exhaustion of moisture sources Insufficient air change rate in HVAC design	Poor vapour barrier installation Poor craftsmanship	Poor indoor ventilation, moisture-intensive activities Poor maintenance of the envelope and system

HVAC, Heating, ventilation and air conditioning.

2015), eventually allowing moisture to accumulate within hygroscopic materials, which, over time, may lead to critical humidity levels (Hens, 2013). In the past, accidental air infiltration assured an adequate ventilation rate, as most of the moisture was transported outside through cracks and holes in the envelope. However, new efficiency standards are creating a push towards sealed and airtight envelopes, thus reducing the opportunity for accidental ventilation to occur (Brambilla & Sangiorgio, 2020; Law & Dewsbury, 2018). On one hand, sealed envelopes reduce uncontrolled air movements between the interior and exterior environments, thus minimising thermal loads and increasing the overall energy efficiency; on the other hand, if not coupled with an adequate mechanical ventilation system, they put the onus on building occupants to properly ventilate the indoor environment. It must be noted that, albeit on a building level, cracks and air infiltrations contribute to reducing the accumulation of moisture in the indoor environment, thus reducing the probability of condensation due to this mechanism; on a component level, uncontrolled air through the envelope may increase the risk of interstitial condensation due to convection.

Some studies suggest that there is a correlation between increasing levels of airtightness and lower air change rates (Crump, Dengel, & Swainson, 2009; Kraus, 2016). Assumptions about window opening behaviour that underpin the assessment of the risk of condensation, moisture-related issues or the hygrothermal indoor environment during the design stage tend to be overestimated (Fabi, Andersen, Corgnati, & Olesen, 2012). This, coupled with the fact that heating and cooling systems in residential buildings do not usually have dehumidification, may explain the higher hygric loads detected in new buildings (Brambilla & Sangiorgio, 2020).

Another construction trend that has completely changed the hygrothermal performance of a building involves the widespread use of lightweight construction, as opposed to massive and heavyweight traditional envelopes. Lightweight envelopes usually have a significantly lower capacity to exchange moisture with the environment, which not only reduces the overall moisture buffering provided but also decreases the hygroscopic range of the material; as a result, pore saturation is reached much faster and condensation is more likely to occur. For example, a traditional brickwork may absorb 1.100 L of water, a timber frame up to 150 and a metallic frame even less (Hens, 2013). This is associated with the extensive use of materials characterised by low vapour permeability, which act as vapour retardants or barriers and further reduce the envelope's drying potential.

On the other hand, new hygroscopic bio-based materials are entering the market, with increased use of timber, straw and sheep wool, among others (Spiegel & Meadows, 2010), that are highly sensitive to moisture (Hoang, Kinney, Corsi, & Szaniszlo, 2010). For this reason, additional protective measures are usually taken by wrapping the material with vapour-resistant foil. However, the environmental benefits of these materials are often prioritised over the effects of their use on overall hygrothermal performance, leading to long-term condensation risks that are often difficult to predict and manage at the design stage.

3.4 Condensation risk assessment

Although hygrothermal assessment and condensation risk predictions are essential to guarantee durability and avoid moisture-related issues, these are not easy tasks. Multiple factors contribute to the complexity of the calculation process and hence hinder its applicability during the design stage. First, moisture transfer is seldom steady state and, unlike heat transfer, is difficult to accurately describe and predict using a simplified approach. The transient nature of moisture transfer is magnified by the fact that materials' hygrothermal properties directly depend on their moisture content, making the problem nonlinear (Rode, 1992). Second, moisture transfer occurs through different processes, including diffusion, liquid transport and convection, making it very difficult to create an easy-to-solve mathematical model that incorporates all these mechanisms (refer to Chapter 6: Hygrothermal modelling). In addition, heat and moisture transfer are closely correlated; they cannot be solved separately but require a simultaneous approach to the problem. Finally, the boundary conditions and the underlying assumptions that must be considered in the assessment are still not well identified in an agreed standard, nor is there a comprehensive body of knowledge that can guide their definition (Rode, 1992).

Steady-state assessment methods are not suitable for accurately predicting condensation risk but are currently the most viable option for designers. Although transient models and computer software for simultaneous heat and moisture calculation are available on the market, they are mainly used in research, due to the requirement for highly precise input that is often not available at the design stage and the considerable level of complexity and expertise involved.

3.4.1 Steady-state methods

Despite the known deficiencies of steady-state assessment methods (White, 1989), most building codes and standards continue to recommend them (refer to Chapter 7: Building codes and standards). One reason for this is that they allow the calculation to be performed without the need for specialised software. Steady-state methods can be used to assess the risk of condensation both on the surface and within the building envelope.

3.4.1.1 Surface condensation: critical surface humidity

Indoor surface condensation occurs when the temperature at that surface is equal to or below the air dew point temperature. When an object has a surface temperature lower than the dew point of the surrounding air, the moisture contained in the layer of air in contact with the surface will condense to equilibrate the partial vapour pressure to the saturation vapour pressure at that temperature. In buildings, condensation occurs on those surfaces where the temperature difference from the indoor air is high due to their lower thermal resistance, such as windows, poorly insulated components or in the presence of thermal bridges, such as studs in frame walls.

The critical surface humidity method can be used to account for the risk of surface condensation during the design stage (EN ISO 13788, 2012). This method indicates that the monthly surface humidity should never exceed the critical surface humidity value, usually considered equal to 0.8, identified as the value that leads to mould growth. The surface humidity depends on the indoor air vapour pressure and the surface temperature of the building component, which is closely correlated with its thermal transmittance. Hence, the method focuses on the overall thermal performance of the building component as a strategy to control the surface temperature and guarantee that the critical humidity threshold is not exceeded. During the design stage, the method recommends a calibration of the overall thermal resistance based on the definition of a parameter, called the temperature factor at the internal surface $f_{R,\ si}$. This expresses the overall thermal quality of a building component and is calculated as:

$$f_{R,si} = \frac{(\theta_{s,i} - \theta_{\text{ext}})}{(\theta_{\text{op}} - \theta_{\text{ext}})} \tag{3.1}$$

where θ_{ext} is the external temperature, $\theta_{s,i}$ the temperature of the internal surface, and θ_{op} the internal operative temperature, all express in °C. Note that, in this formula, temperature is identified with the symbol θ instead of T (as used in Chapter 2: Principles of hygrothermal processes) to keep consistency with the standard referenced.

The design process follows a step-by-step approach:

1. Definition of the monthly external climatic condition (temperature and humidity). External condition should be representative of the building's location, taking account of altitude where appropriate. The values should be taken from the relevant building code or, if these are not available, from the climatic record, where monthly mean value is taken as the one that is likely to occur every 10 years.
2. Definition of the internal temperature based on the relevant building code or national practice.
3. Calculation of the internal relative humidity. If the monthly HVAC is not known and there is no air conditioning system to keep it constant over the year, the value can be calculated as:

$$p_i = p_{\text{ext}} + (\Delta v \cdot R_v \cdot \theta_i) = p_{\text{ext}} + \frac{G}{n \cdot V} R_v \cdot \theta_i$$

where P_i is the partial vapour pressure of the internal air, P_{ext} is the partial vapour pressure of the external air, R_v is the gas constant for vapour, θ is the temperature and G the moisture production.
4. Calculation of the minimum acceptable saturation vapour pressure at the surface based on the value identified as critical humidity at the surface: $P_{\text{sat}} = \frac{P_i}{\text{RH}_{\text{crit}}}$.
5. Calculation of the minimum acceptable surface temperature based on (4).
6. Calculation of the minimum temperature factor $f_{Rsi,\min}$ for each month using (5) and (1).
7. Identification of the month with the highest temperature factor, which is then taken as the critical threshold.
8. Design of the building component's thermal resistance so that the critical factor is always exceeded.

The temperature factor, together with the linear and point thermal transmittance, is also used as a simplified method to assess the risk of condensation in the presence of thermal bridges (Plessis, Filfli, Muresan, & Bouia, 2011). Although there are databases of temperature values associated with common thermal bridges, they are not comprehensive due to the multitude of architectural detailing solutions and the great diversity in construction methods (Bliuc et al., 2017). In this case, the reduction of overall thermal transmittance due to the thermal bridge is accounted for in the calculation of the surface temperature. However, it must be noted that the accuracy of this simplified method ranges from 20% to 50%, due to the dependence of the linear and point thermal transmittance on geometrical, physical, technological and construction factors (Bliuc et al., 2017).

3.4.1.2 Interstitial condensation

Interstitial condensation can occur when moisture accumulates within the material's pores or when vapour migrating through the envelope meets layers at a temperature that is lower than the dew point. Sometimes, the quantity of liquid formed during the process is small and evaporates naturally in a short period of time. In this case, condensation may not be an issue from a durability perspective. In other cases, the limited evaporation potential reduces the drying ability, leading to significant destructive consequences over time.

All interstitial condensation assessment methods compare the partial vapour pressure with the saturation vapour pressure to determine the risk of condensation. However, they are based on a set of assumptions that limit their applicability and accuracy and lead to inaccurate condensation predictions (White, 1989). The steady-state methods are based on a diffusion-only approach to moisture transport, where any moisture accumulation is determined only by the vapour flowing across the envelope. They also neglect air transport, airflow through the building materials, the effects of solar radiation on external surfaces and moisture storage in the pores. The assessment is made on the cross section, meaning that the calculations assume an indefinite plane geometry and a one-dimensional flow, overlooking the effect of boundary conditions. Further, materials are considered homogenous and isotropic and their properties are kept constant, without acknowledgement of their dependence on moisture content.

The most widely used steady-state approach is the so-called Glaser method (Glaser, 1959), developed to determine the distribution of vapour pressure across a building component. The Glaser method uses a graphic approach, where the monthly course of vapour pressure across the envelope is compared with the values of saturation vapour pressure and condensation occurs when the lines intersect. To make the comparison possible, the material thickness must be drawn proportionally to its vapour resistance, rather than the actual geometrical one.

A modified Glaser method allows the quantification of the amount of accumulated moisture due to interstitial condensation caused by the vapour flow rate through the building element. Due to the nature of the calculation, this has to be considered as an assessment tool rather than an accurate prediction tool. The preparation phase for this calculation aims to define the boundary conditions (temperature and humidity of the interior and exterior air) as well as the hygrothermal properties of the building component, such as the total thermal resistance, the total water diffusion-equivalent air layer thickness (s_d), and the same two values for each of the layers. The boundary conditions are defined using the superficial condensation assessment (steps 1–3), while the water vapour diffusion-equivalent air layer thickness is calculated through the equation:

$$s_d = \mu \cdot s \tag{3.2}$$

where s is the real thickness and μ the vapour resistance factor. This parameter is defined as the thickness of a motionless air layer with the same vapour resistance as the material in question.

The total thermal resistance R and s_d are calculated as:

$$R_{\text{tot}} = R_{si} + \sum_{j=1}^{N} R_j + R_{se} \tag{3.3}$$

$$s_{d,t} = \sum_{j=1}^{N} s_{d,j} \tag{3.4}$$

with R_{si} the resistance of the internal surface, R_{se} the resistance of the external surface, R_j and s_{dj} the resistance and water diffusion-equivalent air layer of thickness of each material from the first to the last (1 to N). If the values are to be calculated at each progressive interface, from 1 to n, the contribution of the materials beyond the one under consideration should not be considered. Based on these values, it is possible to quantify the temperature at each interface between the different materials, through:

$$\theta_n = \theta_{\text{ext}} + \frac{R_n}{R_t}(\theta_i - \theta_{\text{ext}}) \tag{3.5}$$

where θ_n is the temperature at the interface n, θ_{ext} the external temperature, θ_i the internal temperature, R_t the total thermal resistance and R_n the total thermal resistance up to the n interface.

The saturation vapour pressures at the interfaces are then calculated from the temperatures, as discussed in Chapter 2, Principles of hygrothermal processes. Once the saturation vapour pressures at each interface are clear, it is possible to proceed with the graphic-based assessment. The component can be drawn in section, substituting the water vapour diffusion-equivalent air layer thickness for the real thickness of the material. Then, a straight line connecting the different saturation vapour pressures can be traced. The partial vapour pressure line is, instead, drawn straight, connecting the interior and exterior values, calculated from the temperatures. If the two lines do not touch, as in Fig. 3.2, this means that condensation is not detected.

If the lines intersect, this means that condensation occurs in that layer, and the vapour pressure line is corrected by making it tangent to the saturation curve at the point of contact and then straight again for the rest of the construction. This process produces a change in slope of the vapour pressure line, reflecting the difference in

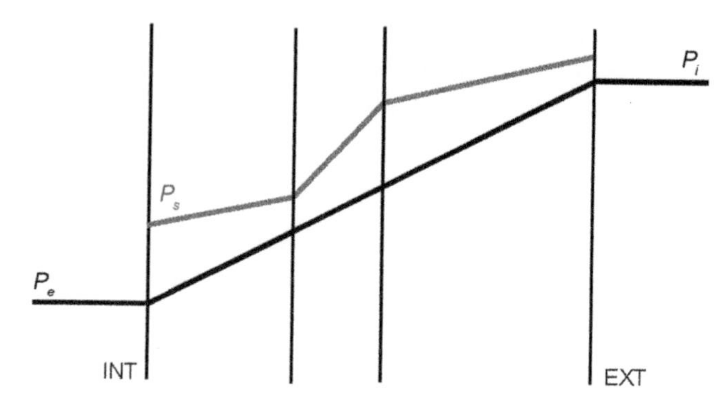

Figure 3.2 The lines that represent the saturation vapour pressure and the partial vapour pressure do not touch.

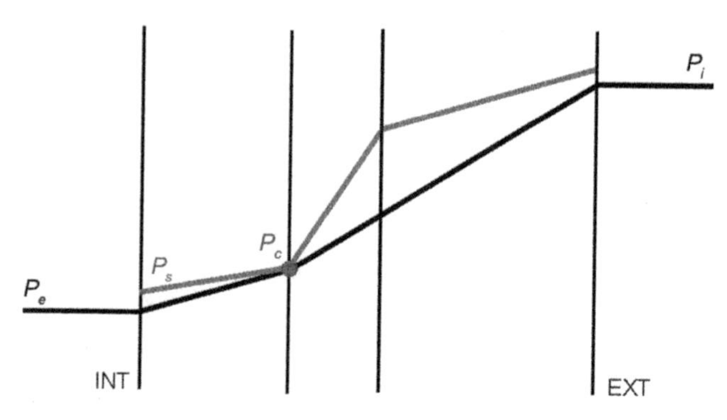

Figure 3.3 Intersection between the lines that represent the saturation vapour pressure and the partial vapour pressure.

the vapour flux before and after the condensation point, which ultimately indicates the amount of vapour condensed. Fig. 3.3 shows the corrected curves.

The calculation must be performed on 12 consecutive months, starting from a precise point. To determine the starting month, the calculation is performed on a trial month, identified as the potential critical one; if no condensation is detected, the process must be repeated for the consecutive months. If no condensation occurs in all 12 months, the likelihood of condensation can be considered very low. If a month shows condensation, it becomes the starting point. On the other hand, if condensation is found in the trial month, the calculation must be performed for the antecedent months to identify the first month in which condensation occurs. This elaborate process ensures that the condensation and evaporation balance over the year is quantified correctly.

The principle of mass conservation can be used to determine the amount of condensation. In equilibrium, the amount of moisture transported from and to a material is equal. Hence, if the two values differ, the difference identifies the amount of moisture that condenses. To calculate the moisture to and from a material, Fick's law can be used in the layers just before and after the condensation interface, as the vapour transfer relies only on diffusion. Therefore the rate of condensation is defined as:

$$g_c = g_{\text{to}} - g_{\text{from}} = \delta_0 \left(\frac{P_i - P_c}{s_{d,t} - s_{d,c}} - \frac{P_c - P_e}{s_{d,c}} \right) \quad (3.6)$$

where c is the layer where condensation occurs, i is the material before, e the material after. If the lines intersect at more than one interface, the condensation rate should be assessed for all the interfaces where condensation occurs.

The equation can also be used to calculate the rate of evaporation, whereas the value g_c will result negative. The rate of condensation and evaporation are then compared to determine whether the building component accumulates moisture or dries during the year. If condensation at one or more interfaces does not completely evaporate, the structure has failed the assessment and needs to be redesigned and tested again.

It is worth noting that, when a vapour barrier is introduced into the construction, the method still applies. In this case, the total thermal resistance does not change, as the thickness of the barrier makes its thermal contribution negligible; hence, the temperature distribution and, consequently, the saturation vapour pressure do not change. However, the slope of the line that represents the partial vapour pressure distribution is modified based on the additional vapour resistance.

Glaser can also be used to assess the envelope's drying potential. The procedure assumes wet materials, simulating the conditions where material has built-in moisture due to the construction process and where eventual leaks from services, defects in the waterproof layer or a previous interstitial condensation problem have been rectified. Hence, an additional moisture content of 1 kg/m^2 is applied at the centre of the layer, while Glaser is used to evaluate the evaporation rate for each month. The calculation is performed measuring the length of time it takes to remove all condensation, and there are three possible outcomes:

1. Drying out within 10 years without condensation in other layers. Depending on the time it takes for the water to evaporate, assess the risk of degradation to the layer containing the excess moisture.
2. Drying out within 10 years with temporary condensation in other layers. Depending on the time it takes for the water to evaporate, assess the risk of degradation to the layer containing the excess moisture and to all other layers that show condensation issues.
3. Drying time exceeds 10 years.

3.4.2 Transient methods

Full models can be used to assess the transient hygrothermal performance of building components. These models usually take account of the complex transport mechanics, heat and moisture storage in building materials, latent heat transfer, moisture-dependent materials' properties, vapour diffusion, liquid transport (surface diffusion and capillary flow), sorption and capillary forces. However, this level of complexity requires computer software and a simulation programme able to solve the heat and moisture transfer simultaneously. Chapter 6, Hygrothermal modelling, discusses the currently available models and programs that can be used for transient hygrothermal assessments.

3.5 Moisture-safe buildings

Condensation and moisture-related risks depend on several factors that interact with each other in a way that is seldom predictable. For this reason, a moisture-safe approach to building design, construction and operation should aim both to reduce – or, ideally eliminate – the likelihood of condensation occurring and to increase the building's resilience in relation to hygrothermal performance. This, however, is a multifaceted issue that needs to be tackled using a holistic approach aimed at improving all the phases and processes that combine to determine the final building's response to moisture.

Because moisture is an element that cannot be removed from the equation, the strategies that can be implemented should aim at minimising and controlling the risk factors. These strategies can be grouped into three main categories: process, design and construction, and preservation.

1. The first group relates to the holistic multidisciplinary approach, from policy to practice. Currently, there is little, if any, knowledge exchange among the various actors involved in the whole life cycle of a building. This is clearly an impediment to the possibility of integrating long-term strategies and approaches to manage risk over time.
2. The second group includes all the choices and practices undertaken during the design and construction stages, from the architectural concept to material selection and installation. The lack of an overarching prevention agenda often produces a cascading effect on the different processes in the early stages of a building's life, forcing each actor to consider moisture safety only in one limited domain, if it is considered at all. This results in a constellation of separate, fragmented strategies that may conflict with one another and nullify the various efforts.
3. Preservation strategies involve occupants' behaviour as well as maintenance and repair practices. A moisture-safety approach should aim to minimise those activities that increase the risk of condensation and ensure the existence and correct operation of an integrated plan to achieve optimal hygrothermal performance over time.

3.5.1 *Improving the process*

Although there is a body of knowledge about condensation risks, effects and causes, the building sector is compartmentalised into separate domains. As a result, each actor involved in the process deals with only one domain and cannot see the bigger picture. This aggravates moisture-related issues, which require a thorough and integrated approach from the early stages of design to the long-term occupation and preservation phase. The fragmentation of the sector is apparent on multiple levels. For example, building occupants might not be aware of the intended use of the building

Table 3.3 Suggestion for the improvement of the process to ensure condensation-free and durable constructions.

Area	Strategies
General policy	• Develop consistent criteria for moisture management • Improve the building codes in regard to maintenance, operation and design, based on up-to-date knowledge • Increase opportunity and milestone in the process to exchange knowledge and experience between the different stakeholders
Design	• Identification of critical work, process and stages in regard to moisture management • Ensure compliancy with up-to-date documentation on moisture safety in design
Construction	• Ensure coordination with the design phase in regard to documentations, drawings and specification of the critical works and detailing • Recording construction and detailing of critical processes
Preservation	• Keep and updating the preservation plan • Develop occupation guidelines • Develop a shared postoccupancy evaluation procedure

and building systems that was assumed during the design stage and which drove the performance assessment, such as the opening of windows and ventilation behaviour. Similarly, a designer is not necessarily aware of the future occupants' habits. Building codes and industry handbooks do not provide a solution to this problem, and none of the existing documents addresses the integration of overarching strategies that could ensure moisture safety from design to operation (Mjornell et al., 2012).

To ensure condensation-free and durable constructions, it is of utmost importance to establish a thorough workflow plan with clear responsibilities and cross-referenced quality checks across the design, construction and occupation phases. Table 3.3 provides some suggestions for improvement in the process (Australian Building Code Board, 2014b; Franchi et al., 2004; Mjornell et al., 2012).

The holistic approach to the whole building life cycle represents best practice in moisture-safety design, as it ensures an integrated process in which each stakeholder is aware of the implications of their actions and choices for the overall long-term hygrothermal performance. Exchange of knowledge and experience, as well as systematic documentation and communication across all stages of building design, construction and operation, is of particular importance to guarantee an effective workflow that bridges the existing gaps in the sector.

3.5.2 Improving the design and construction phases

The design phase is critical for moisture safety, as it lays the foundation for the success of all the other phases: construction, maintenance and occupation. For this reason, condensation prevention starts with the design approach, which should aim to identify the risk factors for the specific case, understand how it is possible to reduce the likelihood of condensation, and establish appropriate management strategies for the phases that follow.

3.5.2.1 Whole building approach

If the first step is the identification of the risk factors influencing condensation in the specific context of the project, as well as the critical phases, activities and processes that might increase the risk, the second phase targets the overall design approach. The risk of condensation can be minimised by implementing an approach focused on the simultaneous control of moisture entry and accumulation, as the strategies implemented for one may not work for, or be detrimental to, the other. Strategies that are effective in limiting moisture infiltration are likely to be equally effective in preventing moisture escaping from the envelope. As a general indication, it is possible to identify a design workflow aimed at, first, preventing the envelope getting wet from outdoor moisture sources; second, preventing the envelope getting wet from indoor moisture sources; and finally, ensuring opportunities to dry.

The major transport mechanisms relevant to outdoor moisture sources are liquid flow and capillary suction from rain, dew and groundwater (Lstiburek, 2002). Hence, shedding of water and protection from rain, as well as capillary breaks on the footings, should be considered to reduce possible moisture pressure on the envelope. Rainwater leaks through joints between different parts of the building envelope are extremely common, posing a significant risk of condensation. It is of utmost importance that the overall design approach accounts for this by providing resilience to failure as well as redundancy to ensure performance in case a failure does occur.

Airflows of moisture-laden air and infiltrations are the second most important wetting mechanisms (Lstiburek, 2002). These movements are usually driven by differential vapour pressure, which is related to indoor and outdoor conditions such as moisture content or relative humidity. Accordingly, strategies and techniques to manage air movements and vapour diffusion across the building envelope must be carefully selected based on the context. Generally, this is reflected in strategies that prevent exfiltration in cold climates and infiltration in warm climates. At the whole building level, this can be achieved through the integrated design of the mechanical systems, which can assist via slight pressurisation and depressurisation of the indoor spaces. At the same time, careful consideration must be given to component design and the permeability of the selected barriers in the development of a holistic and comprehensive risk reduction strategy.

Excessive indoor moisture also presents a significant condensation risk, especially when coupled with airtight and sealed envelopes that reduce the opportunity for uncontrolled ventilation (Brambilla & Sangiorgio, 2020). Hence, it is important to introduce strategies aimed at the management of the indoor environment from the early design stage. Given that occupants' behaviour is unlikely to be predictable or controllable, attention must switch to the provision of robust systems that guarantee a certain level of resilience to nonoptimal use of the spaces. Mechanical heating, cooling and ventilation systems, for example, can be used to maintain indoor humidity within safe ranges. Ventilation of the indoors is also critical in moisture management (Winkler, Munk, & Woods, 2018). Therefore hybrid systems should be preferred to natural ventilation to achieve appropriate ventilation rates and improve control over dehumidification, heat recovery and air filtration (Brambilla and Sangiorgio, 2020). The success of these strategies is, however, closely related to the operational phase: occupants' activities and regular inspections are essential to guarantee long-term performance (Seppänen, 2004).

3.5.2.2 Component design and material selection

Component design and material selection are delicate phases in the moisture-safety design process, as they determine the envelope's hygrothermal response and, ultimately, the risk of interstitial condensation. Even if moisture sources and climatic risk factors are minimised, inappropriate envelope strategies can still lead to condensation. The first step in component design is function analysis, where the functions are described and detailed in regard to the relevant performance requirements (Soronis, 1992). This phase identifies the requirements and limitations that should guide the selection of the materials.

Using function analysis, it is possible to define the moisture control strategies suitable for the specific case, which should target the prevention of moisture accumulation as well as the provision of drying opportunities. Building materials are generally hygroscopic, allowing for diffusion to take place; this is not hazardous as long as the moisture stored within the pores is allowed to dry. There are three main strategies for dealing with the drying capacity of building components.

The first strategy focuses on exterior moisture sources. It allows drying towards the outside and reduces exchanges towards the interior by placing a moisture barrier on the interior side of the insulation layer and a breathable weather membrane on the exterior side. This component typology is particularly common in cold climates, where moisture flux usually occurs towards the outside and interstitial condensation is likely to occur on the insulation. In this case, it is essential to prevent the envelope getting wet from indoor moisture sources while allowing a degree of permeability on the exterior surface.

The second strategy relies on the opposite concept, allowing drying towards the interior. This is achieved by installing a permeable membrane on the interior side and a vapour-resistant one on the exterior side. This typology is more suitable for warm climates, where moisture flux occurs in the opposite direction, from outdoor to indoor space, usually at lower temperature and humidity levels due to the air conditioning system.

The third strategy takes a completely different approach by creating a totally breathable component. This 'flow-through' approach may be suitable for mixed continental climates, where the moisture flux changes direction periodically, according to seasonal or daily variations. However, it requires both air pressure and indoor humidity control to avoid moisture accumulation.

The strategy adopted for the component design determines the materials that should be used to create a coherent vapour control strategy. It is important to install hygroscopic materials on the side where drying and wetting are allowed, which would otherwise act as a second vapour-resistant barrier and create a completely sealed envelope with no opportunity for evaporation. Further, materials should be specified in regard to their condition of use: locations with cyclical, frequent and recurring high moisture loads, such as bathrooms, should be clad with low-hygroscopic materials able to withstand repeated wetting (WHO, 2009). In this context, it is worth noting that innovative bio-based materials that are moisture-sensitive or which have not been tested over a long period may constitute a liability in the overall design strategy.

3.5.2.2.1 Vapour barriers and permeability

Although vapour barriers are essential for a comprehensive moisture control strategy, these elements can also increase the risk of condensation if not appropriately designed. Vapour barriers should be designed to manage the conflicting requirements of controlling vapour flow by encouraging drying as well as preventing wetting. The following principles can be applied to manage this complexity:

- prefer vapour retarders to vapour barriers;
- avoid vapour barriers on both sides of the building component; drying should be allowed in at least one direction;
- avoid interior vapour barriers in spaces that are fully air conditioned throughout the year; and
- avoid impermeable interior claddings if not required.

These principles reflect the difference between vapour retarders and vapour barriers. The former are materials that allow a certain amount of vapour to pass through, while the latter are completely impermeable to vapour. Membranes are classified according to their permeability, which usually ranges from less than 0.1 (e.g. sheet polyethylene, unperforated aluminium foil) to 10 perm (e.g. latex or enamel paint appropriately rated and installed).

3.5.2.3 Construction

All the design strategies aimed at minimising the risk of condensation must be followed by equal attention to moisture issues during the construction phase. This involves care in the installation, storage and management of the components, materials and systems. Indeed, the way materials are stored on the construction site is just as important as the installation and construction phase to the success of moisture-management strategies. Hygroscopic materials should be placed in protected locations, isolated from the ground and not exposed to the environment to minimise exposure to construction-related moisture sources.

3.5.2.4 Practical strategies to minimise condensation risk

Table 3.4 presents practical strategies that can be used to reduce the risk of condensation during the design and construction phase (Mjornell et al., 2012; Roberts, 2008; Sanders & Phillipson, 2003). The list is not meant to be exhaustive.

Table 3.4 Practical strategies to minimise condensation risk during the design and construction phase.

Area	Strategies
Design strategies	• Design and specify flashing and protective layer connections • Increase ventilation rate through hybrid mechanical systems • Increase control over the indoor environment through HVAC with dehumidification • Reduce thermal bridges by designing continuous external insulation • Venting fans outlet outdoor and not in the roof attic
Technological solutions	• Increase ventilation of roof and underfloor cavities • Use low-hygroscopic materials in moisture-intensive location • Select materials based on location and use • Replace carpet with less hygroscopic materials • Use materials that can withstand cyclical moisture loads in bathrooms, kitchens, below-grade floors • Use air barrier to prevent air infiltration • Add additional weep-hole sin cavity walls to increase drying opportunity • Increase drainage capacity through perforate vertical battens • In cold climate locate the vapour barrier towards the interior • In warm climate on the exterior side
Construction	• Store materials in a ventilated and protected area • Monitor environmental conditions in storage spaces • Dispose materials that have been stored improperly and became damp • Protect timber elements from weather during and after installation • Avoid direct contact of lining plasterboard with concrete or creed • Avoid installing mineral wool insulation in contact with wet layers • Install a vapour barrier between timber and concrete elements • Install weather seals and insulation on ducts and pipes • Allow the cast-in-place concrete to dry before installing the envelope • In the last part of the construction process, use the mechanical system to dehumidify the internal air and remove the moisture and water trapped indoor during the construction process

HVAC, Heating, ventilation and air conditioning.

3.5.3 Improving the prevention plan and occupational practices

Design and construction strategies aimed at preventing condensation and moisture-related durability are necessary to minimise the risks but are not sufficient on their own. The practices of building occupants and managers determine the final hygrothermal performance.

Table 3.5 Strategies to improve the occupation phase, including occupants' practices and management plan.

Area	Strategies
Occupants	• Monitor indoor environmental conditions • Report of failures and damages related to moisture and water • Increase air change ventilation rate
Repair and maintenance	• Incorporate a moisture control plan and regularly update it • Increase the frequency of inspections • Plan periodical audit of indoor occupancy habits in regard to moisture-intensive activities • Immediate replacement of damp and damaged components

Occupants' behaviour is hardly predictable but is well acknowledged as a major factor in indoor moisture accumulation. As discussed previously, moisture-intense activities can magnify the condensation risk, especially when coupled with under-ventilation of the spaces. Activities that increase indoor humidity should be minimised and those that support the reduction of moisture loads should be encouraged. Although the correct management of the indoor environment is the responsibility of occupants, it is also essential to raise awareness and promote appropriate behaviour in regard to moisture control (Brambilla & Sangiorgio, 2020). A comprehensive preservation plan should include educational activities, guidelines and suggestions designed to inform occupants about their role in helping to reduce condensation risks. In addition, periodical audits of indoor moisture generation levels and a continuous indoor environmental monitoring campaign can inform modifications of practices and habits (Australian Building Code Board, 2014b; Franchi et al., 2004; Mjornell et al., 2012). Frequent inspections and detailed maintenance programs are also necessary to provide a complete picture by assessing the impact on the building envelope, components and materials (Adams et al., 2016; Sanders & Phillipson, 2003; Seppanen, Fisk, & Mendell, 1999). The early detection of failures, malfunctions or defects can significantly reduce the impact of condensation and help to prevent its occurrence over the long term (Seppänen, 2004).

Table 3.5 presents practical strategies that can be used to reduce the risk of condensation during the occupation phase (Fisk, Mirer, & Mendell, 2009; Menzies & Bourbeau, 1997; Wargocki et al., 2002), it includes occupants' habits and management practices (Australian Building Code Board, 2014b; Franchi et al., 2004; Mjornell et al., 2012).

3.6 Durability in energy efficiency codes

In recent years, with increasing awareness of the limitations of a vision of sustainability merely based on energy efficiency, green building rating tools have begun to consider durability as an essential feature of sustainable buildings. However, it is an extremely complex task to define a durable building and determine a set of expectations in relation to lifespan and performance.

Despite the importance of condensation as a factor in durability, health and well-being, as well as sustainability, only a small number of standards and rating tools clearly and directly address this dimension. Often, a condensation assessment is not prescribed or required but is left to the designer's discretion. Other approaches included in building codes and standards are discussed in Chapter 7, Building codes and standards. Some energy performance rating tools address condensation indirectly via limitations and requirements on aspects that influence durability. For example, Passive House focuses on providing a comfortable internal environment and on the improvement of the overall thermal performance of the envelope by increasing the total thermal resistance and airtightness and minimising thermal bridges. This rating scheme requires a check of the factors associated with surface temperatures ($f_{R,si}$) to tackle the risk of temperature asymmetry and surface condensation at the presence of thermal bridges (Passive House, n.d.).

However, most green rating tools focus on durability as a holistic concept that involves the whole design process, rewarding those projects that are able to provide evidence of long-term performance (Reeder, 2010). For example, the BREEAM rating system includes durability performance requirements in the Materials category (Mat 05 – Designing for durability and resilience) based on the preservation plan. BREEAM requires, 'The relevant building elements incorporate appropriate design and specification measures to limit material degradation due to environmental factors'. To demonstrate fulfilment of the criteria it is necessary to (1) identify the critical building components, (2) identify the environmental risk factors, (3) demonstrate that the design and specifications account for these degradation factors and (4) involve an external assessor to review the design and construction phase and assure that specifications are followed (BREEAM, 2014).

LEED, on the other hand, includes a Materials and Resources Credit – Durable Building, which is awarded to projects that integrate a preservation plan and demonstrate that the predicted service life of the building components is greater than the design service life. To satisfy the requirements it is necessary to form a dedicated workforce, called a Design Service Team, which must provide evidence that the design has followed specific criteria to account for durability issues. For example, the taskforce needs to quantify the design service life of each building component and provide an extensive preservation document that lists the risks and the maintenance schedule, including an indication of frequency and costs, for each building subsystem (*LEED*, n.d.).

However, these tools tend to overlook the influence of the building's occupants. While they provide guidance on how to build more sustainably during the design and construction stages, an integrated design approach to durability should also account for the occupational phase. LEED includes credit for Awareness and Education, requiring that owners receive training and information on how to operate the building and manage the maintenance to ensure that its requirements are met over time. In contrast, in the Living Building Challenge's credit system, a project is only certified after 1 year of operation, including the building occupants' practices.

3.7 Concluding remarks

The close correlation between design, construction, occupation and management of buildings highlights the complexity associated with durability and condensation prevention. New trends and practices change the hygrothermal response of the building envelope, making it even more difficult to provide consistent and robust strategies that could improve overall performance and reduce the condensation risk. This complexity calls for the development of a coherent overarching strategy that includes all the stakeholders involved in the correct management of the risk across the whole building life cycle. The first step should involve an update of the current policy framework to fill the gaps in the process that lead to fragmentation and disconnection from one phase to another, develop a consistent approach and promote communication and exchange of knowledge among the various stakeholders.

A multidisciplinary research approach is needed to advance the understanding and improve the building design and maintenance practices by integrating the expertise, knowledge and methods from different fields. Indeed, this issue requires a comprehensive and holistic research programme that could bridge construction, design and health, while informing policy and human behaviour (Brambilla & Sangiorgio, 2020).

References

Adams, R. I., Bhangar, S., Dannemiller, K. C., Eisen, J. A., Fierer, N., Gilbert, J. A., ... Bibby, K. (2016). Ten questions concerning the microbiomes of buildings. *Building and Environment, 109*, 224–234. Available from https://doi.org/10.1016/j.buildenv.2016.09.001.

Aisner, J., Schimpff, S. C., Bennett, J. E., Young, V. M., & Wiernik, P. H. (1976). *Aspergillus* infections in cancer patients: Association with fireproofing materials in a new hospital. *Journal of the American Medical Association, 235*(4), 411–412. Available from https://doi.org/10.1001/jama.235.4.411.

Allsopp, D., Seal, K. J., & Gaylarde, C. C. (2004). *Introduction to biodeterioration*. Cambridge University Press.

Ashton, H. E., & Sereda, P. J. (1981). *Environment, microenvironment, and durability of building materials*. (pp. 28–31). NBS.

Australian Building Code Board. (2014a). *Australian Building Code*.

Australian Building Code Board. (2014b). *Handbook: Condensation in buildings*.

Australian Building Code Board. (2015). *Handbook: Durability in buildings, including plumbing installations*.

Bliuc, I., Lepadatu, D., Iacob, A., Judele, L., & Bucur, R. D. (2017). Assessment of thermal bridges effect on energy performance and condensation risk in buildings using DoE and RSM methods. *European Journal of Environmental and Civil Engineering, 21*(12), 1466–1484. Available from https://doi.org/10.1080/19648189.2016.1172032.

Brambilla, A., & Sangiorgio, A. (2020). Mould growth in energy efficient buildings: Causes, health implications and strategies to mitigate the risk. *Renewable and Sustainable Energy Reviews, 132*, 110093. Available from https://doi.org/10.1016/j.rser.2020.110093.

BREEAM. (2014). *Mat 05 – Designing for durability and resilience*. https://www.breeam.com/BREEAMUK2014SchemeDocument/content/09_material/mat05.htm.

BS. (2002). Code of practice for control of condensation in buildings (p. 5250).

Cai, S., Zhang, B., & Cremaschi, L. (2017). Review of moisture behavior and thermal performance of polystyrene insulation in building applications. *Building and Environment, 123*, 50–65. Available from https://doi.org/10.1016/j.buildenv.2017.06.034.

Calle, K., Coupillie, C., Janssens, A., & Van Den Bossche, N. (2020). Implementation of rainwater infiltration measurements in hygrothermal modelling of non-insulated brick cavity walls. *Journal of Building Physics, 43*, 477–502. Available from https://doi.org/10.1177/1744259119883909.

Crump, D., Dengel, A., & Swainson, M. (2009). *Indoor air quality in highly energy efficient homes—A review*. Buckinghamshire: NHBC Foundation.

Derome, D. (1999). Moisture occurrence in roof assemblies containing moisture storing insulation and its impact on the durability of building envelope (Doctoral dissertation, Concordia University).

Dewsbury, M., Law, T., & Henderson, A. (2016). *Investigation of destructive condensation in Australian cool-temperate buildings*. DoJ Building Standards and Occupational Licensing, Editor.

EN ISO 13788. (2012). *Hygrothermal performance of building components and building elements—Internal surface temperature to avoid critical surface humidity and interstitial condensation—Calculation methods*.

Fabi, V., Andersen, R. V., Corgnati, S., & Olesen, B. W. (2012). Occupants' window opening behaviour: A literature review of factors influencing occupant behaviour and models. *Building and Environment, 58*, 188–198. Available from https://doi.org/10.1016/j.buildenv.2012.07.009.

Fantucci, S., Isaia, F., Serra, V., & Dutto, M. (2017). Insulating coat to prevent mold growth in thermal bridges. *Energy Procedia, 134*, 414–422. Available from https://doi.org/10.1016/j.egypro.2017.09.591.

Fisk, W. J., Mirer, A. G., & Mendell, M. J. (2009). Quantitative relationship of sick building syndrome symptoms with ventilation rates. *Indoor Air, 19*, 159–165. Available from https://doi.org/10.1111/j.1600-0668.2008.00575.x.

Franchi, M., Carrer, P., Kotzias, D., Rameckers, E. M., Seppänen, O., van Bronswijk., ... Viegi, G. (2004). *Towards Healthy Air in Dwellings in Europe*. Brussels: European Federation of Allergy and Airways Diseases Patients Associations.

Glaser, H. (1959). Graphisches verfahren zur untersuchung von diffusionsvorgängen (p. 11).

Hanafi, M. H., Umar, M. U., Razak, A. A., Rashid, Z. Z. A., Noriman, N. Z., & Dahham, O. S. (2018). An introduction to thermal bridge assessment and mould risk at dampness surface for heritage building. In *In IOP Conference Series. Materials Science and Engineering*, (454, p. 012185). IOP Publishing.

Hao, L., Herrera, D., Troi, A., Petitta, M., Matiu, M., & Del Pero, C. (2020). Assessing the impact of climate change on energy retrofit of alpine historic buildings: Consequences for the hygrothermal performance. *IOP Conference Series: Earth and Environmental Science, 410*, 012050. Available from https://doi.org/10.1088/1755-1315/410/1/012050.

Hens, H. L. (2013). Moisture control in buildings. In *A Handbook of Sustainable Building Design and Engineering: An Integrated Approach to Energy, Health and Operational Performance*, 211.

Hens, H. (2017). *Building physics—Heat, air and moisture: Fundamentals and engineering methods with examples and exercises*.

Hoang, C. P., Kinney, K. A., Corsi, R. L., & Szaniszlo, P. J. (2010). Resistance of green building materials to fungal growth. *International Biodeterioration and Biodegradation, 64*(2), 104–113. Available from https://doi.org/10.1016/j.ibiod.2009.11.001.

Ilomets, S., & Kalamees, T. (2016). Evaluation of the criticality of thermal bridges. *Journal of Building Pathology and Rehabilitation, 1*. Available from https://doi.org/10.1007/s41024-016-0005-6.

Ilomets, S., Kalamees, T., & Vinha, J. (2018). Indoor hygrothermal loads for the deterministic and stochastic design of the building envelope for dwellings in cold climates. *Journal of Building Physics, 41*(6), 547–577. Available from https://doi.org/10.1177/1744259117718442.

ISO 14040. (1997). *Environmental management, life cycle assessment: Principles and framework.*

Johansson, P., Pallin, S., & Shahriari, M. (2010). *Risk assessment model applied on building physics: Statistical data acquisition and stochastic modeling of indoor moisture supply in Swedish multi-family dwellings.* In IEA Annex 55 RAP-RETRO.

Kalamees, T., Vinha, J., & Kurnitski, J. (2006). Indoor humidity loads and moisture production in lightweight timber-frame detached houses. *Journal of Building Physics, 29*(3), 219–246. Available from https://doi.org/10.1177/1744259106060439.

Keskikuru, T., Salo, J., Huttunen, P., Kokotti, H., Hyttinen, M., Halonen, R., & Vinha, J. (2018). Radon, fungal spores and MVOCs reduction in crawl space house: A case study and crawl space development by hygrothermal modelling. *Building and Environment, 138*, 1–10. Available from https://doi.org/10.1016/j.buildenv.2018.04.026.

Kraus, M. (2016). *Airtightness as a key factor of sick building syndrome (SBS), . International multidisciplinary scientific geoconference* (2, pp. 439–445). SGEM: Surveying Geology & Mining Ecology Management.

Lacasse, M. (2003). Durability and performance of building envelopes. In: *BSI 2003 proceedings.*

Law, T., & Dewsbury, M. (2018). The unintended consequence of building sustainably in Australia. In: *Sustainable development research in the Asia-Pacific Region.*

LEED. (n.d.). https://www.usgbc.org/leed.

Lewry, A. J., & Crewdson, L. F. E. (1994). Approaches to testing the durability of materials used in the construction and maintenance of buildings. *Construction and Building Materials, 8*(4), 211–222. Available from https://doi.org/10.1016/S0950-0618(09)90004-6.

Lstiburek, J. (2002). Moisture control for buildings. *ASHRAE Journal, 44*(2), 36–41. Available from https://subscriptions.techstreet.com/subgroups/16359?on_subscription = yes.

Lucchini, A. (1990). Models for the evaluation of the service life of building components. In: *Fifth international conference on durability of buildings materials and components* (pp. 615–624).

Maia, J., Ramos, N. M. M., & Veiga, R. (2019). A new durability assessment methodology of thermal mortars applied in multilayer rendering systems. *Construction and Building Materials, 222*, 654–663. Available from https://doi.org/10.1016/j.conbuildmat.2019.06.178.

Masters, L. W., & Brandt, E. (1989). Systematic methodology for service life prediction of building materials and components. *Materials and Structures*, 385–392. Available from https://doi.org/10.1007/bf02472509.

Matilainen, M., & Kurnitski, J. (2003). Moisture conditions in highly insulated outdoor ventilated crawl spaces in cold climates. *Energy and Buildings*, 175–187. Available from https://doi.org/10.1016/s0378-7788(02)00029-4.

Menzies, D., & Bourbeau, J. (1997). Building-related illnesses. *New England Journal of Medicine*, 1524–1531. Available from https://doi.org/10.1056/NEJM199711203372107.

Mjornell, K., Arfvidsson, J., & Sikander, E. (2012). A method for including moisture safety in the building process. *Indoor and Built Environment, 21*(4), 583–594. Available from https://doi.org/10.1177/1420326X11428340.

Newport Partners & ARES Consulting. (2015). *Durability by design: A professional's guide to durable home design.* U.S. Department of Housing and Urban Development.

Ogniewicz, Y., & Tien, C. L. (1981). Analysis of condensation in porous insulation. *International Journal of Heat and Mass Transfer, 24*(3), 421–429. Available from https://doi.org/10.1016/0017-9310(81)90049-1.

Oppenheim, P., & Brennan, T. (2008). Preventing defect claims in hot, humid climates. *ASHRAE Journal, 50.*

Ortega-Calvo, J. J., Hernandez-Marine, M., & Saiz-Jimenez, C. (1991). Biodeterioration of building materials by cyanobacteria and algae. *International Biodeterioration, 28,* 165–185. Available from https://doi.org/10.1016/0265-3036(91)90041-o.

Pallin, S., Johansson, P., & Hagentoft, C. E. (2011). Stochastic modeling of moisture supply in dwellings based on moisture production and moisture buffering capacity. In: *Proceedings of building simulation 2011: 12th conference of international building performance simulation association* (pp. 366–373). http://www.ibpsa.org/proceedings/BS2011/P_1221.pdf.

Passive House. (n.d.). http://www.passivehouse.com.

Plessis, G., Filfli, S., Muresan, C., & Bouia, H. (2011). Using design of experiments methods to develop low energy building model under Modelica. In: *Proceedings of building simulation 2011: 12th conference of international building performance simulation association* (pp. 988–995). http://www.ibpsa.org/proceedings/BS2011/P_1370.pdf.

Reeder, L. (2010). *Guide to green building rating systems: understanding LEED, Green Globes, Energy Star, the National Green Building Standard, and more* (Vol. 12). John Wiley & Sons.

Roberts, S. (2008). Effects of climate change on the built environment. *Energy Policy, 36,* 4552–4557. Available from https://doi.org/10.1016/j.enpol.2008.09.012.

Rode, P. C. (1992). Prediction of moisture transfer in building constructions. *Building and Environment, 27,* 387–397. Available from https://doi.org/10.1016/0360-1323(92)90038-q.

Sanders, C. H., & Phillipson, M. C. (2003). UK adaptation strategy and technical measures: The impacts of climate change on buildings. *Building Research and Information, 31* (3–4), 210–221. Available from https://doi.org/10.1080/0961321032000097638.

Santos, G., Henrique, N., Mendes, P. C., & Philippi. (2009). A building corner model for hygrothermal performance and mould growth risk analyses. *International Journal of Heat and Mass Transfer, 52,* 4862–4872.

Seppänen, O. (2004). Improvement of indoor environment in European residences to alleviate the symptoms of allergic and asthmatic children and adults. *Towards healthy air in dwellings in Europe. The THADE report. Brussels: European Federation of Allergy and Airways Diseases Patient Associations.*

Seppanen, O. A., Fisk, W. J., & Mendell, M. J. (1999). Association of ventilation rates and CO_2 concentrations with health and other responses in commercial and institutional buildings. *Indoor Air, 9,* 226–252. Available from https://doi.org/10.1111/j.1600-0668.1999.00003.x.

Soronis, G. (1992). The problem of durability in building design. *Construction and Building Materials, 6*(4), 205–211. Available from https://doi.org/10.1016/0950-0618(92)90039-2.

Spiegel, R., & Meadows, D. (2010). *Green building materials: a guide to product selection and specification.* John Wiley & Sons.

Stefanoni, M., Angst Ueli, M., & Elsener, B. (2018). Electrochemistry and capillary condensation theory reveal the mechanism of corrosion in dense porous media. *Scientific Reports*, *8*. Available from https://doi.org/10.1038/s41598-018-25794-x.

Vardoulakis, S., Marks, G., & Abramson, M. J. (2020). Lessons Learned from the Australian Bushfires: Climate Change, Air Pollution, and Public Health. *JAMA Internal Medicine*, *180*(5), 635–636. Available from https://doi.org/10.1001/jamainternmed.2020.0703.

Wargocki, P., Sundell, J., Bischof, W., Brundrett, G., Fanger, P. O., Gyntelberg, F., ... Wouters, P. (2002). Ventilation and health in non-industrial indoor environments: Report from a European Multidisciplinary Scientific Consensus Meeting (EUROVEN). *Indoor Air*, *12*(2), 113–128. Available from https://doi.org/10.1034/j.1600-0668.2002.01145.x.

White, J. H. (1989). Moisture Transport in Walls: Canadian Experience. In *Water Vapor Transmission Through Building Materials and Systems: Mechanisms and Measurement*, ASTM International.

WHO. (2009). *WHO guidelines for indoor air quality: Dampness and mould*. World Health Organization. Available from https://www.euro.who.int/__data/assets/pdf_file/0017/43325/E92645.pdf?ua = 1.

Winkler, J., Munk, J., & Woods, J. (2018). Effect of occupant behavior and air-conditioner controls on humidity in typical and high-efficiency homes. *Energy and Buildings*, 364–378. Available from https://doi.org/10.1016/j.enbuild.2018.01.032.

Yik, F. W. H., Sat, P. S. K., & Niu, J. L. (2004). Moisture generation through Chinese household activities. *Indoor and Built Environment*, *13*(2), 115–131. Available from https://doi.org/10.1177/1420326X04040909.

Zemitis, J., Anatolijs, B., & Frolova, M. (2016). Measurements of moisture production caused by various sources. *Energy and Buildings*, *127*, 884–891.

Health and mould growth

4

4.1 Fungi, mites and bacteria in buildings

The evolution of architecture and construction reflects the socio-cultural transformations in human populations. During the Anthropocene, there was a dramatic shift from a nomadic to a sedentary lifestyle. Today, we spend almost 90% of our lives indoors (Klepeis et al., 2001), and architecture is required to provide an enclosed environment that can protect occupants from external conditions.

In recent years, the push for more energy-efficient constructions and increasing levels of air pollution in large cities has favoured a design that aims to provide air-tight and highly insulated buildings. Case studies show that this focus on the external envelope often led to the neglect of indoor air quality control and management. This fact, coupled with the widespread use of building materials that can easily support microbial growth, has resulted in an increased risk of organic proliferation. Fungi and bacteria germination is favoured by inappropriate design choices, such as unsuitable build-ups, the presence of thermal bridges and air leakage, as well as poor maintenance programmes and operational regimens, along with insufficient indoor ventilation and high-moisture loads (Sedlbauer, 2002; Vereecken & Roels, 2012).

The microfungi that colonise buildings are colloquially called *mould*. Their spores are naturally found in the air and on various materials, and under favourable conditions, they can grow and germinate to form a fungal mass, called *mycelium*. The microbial community that can be found in a building is influenced by the exchange between the outdoor and indoor environments and the occupants (Leung & Lee, 2016), as well as accumulated organic compounds (Viitanen et al., 2010) and dust (Borrego & Perdomo, 2012).

The increase in mould growth in indoor environments is currently a matter of great concern around the globe. It has been estimated that the proportion of buildings affected by mould is as high as 45% in Europe, 40% in the United States, 30% in Canada (Nielsen & Thrane, 2001) and 50% in Australia (WHO, 2009). Mould is responsible for the biodeterioration of building materials (Allsopp, Seal, & Gaylarde, 2004) and is not simply an aesthetic issue. Mould has significant social and economic impacts (Gutarowska & Piotrowska, 2007). For example, in 2001 it was estimated that it cost Germany more than 200 million euros to refurbish mouldy indoor spaces (Sedlbauer, 2001). Furthermore, the health effects of mould on susceptible subjects are consistent with those of other environmental risk factors such as tobacco and air pollutants (Peat, Dickerson, & Li, 1998). This highlights the potential health hazards for building occupants.

Moisture and Buildings. DOI: https://doi.org/10.1016/B978-0-12-821097-0.00004-7

This chapter discusses the classification of microorganisms that infest our buildings (fungi, mites and bacteria), their adverse effects on human health, the main factors that affect their growth and methods of assessment.

Temperature and humidity are considered the two main environmental factors that support microbial growth, together with the substrate on which the spore is deposited. The level of acidity (pH), exposure to light, surface roughness and biotic interactions are all characteristics that must be considered (Krus, Sedlbauer, Zillig, & Künzel, 2001). The various organisms that may be found in damp houses are adapted to survive through cycles of favourable and unfavourable conditions (Ortega-Calvo, Hernandez-Marine, & Sáiz-Jiménez, 1991; Viitanen & Bjurman, 1995), thus increasing the risk of germination when spore diffusion is not obstructed. Damp indoor environments create favourable conditions for fungi and bacteria to grow and proliferate.

4.1.1 Fungi

A fungus is composed of long chains of cells, called hyphae, that branch to form the mycelium, a filamentous mass that can be seen without the need for a microscope. Fungal germination starts with a single spore that finds favourable conditions in which to grow a hypha. These conditions are usually determined by the availability of water or moisture and nutrients.

The first hypha is called the germ tube, and it initiates the proliferation process. The germ tube germinates in different directions, creating several branches that can expand very quickly. Growth usually follows a radial pattern, with hyphae extending in all directions. Consequently, moulds have a large surface to volume ratio, allowing them to maximise the explored area in search of nutrients to sustain the expansion. The hyphae are cross-connected to allow for the efficient movement of nutrients, increasing the growth potential and compensating for those zones where supplements and foods are limited (Fig. 4.1).

Grown fungi produce spores that are designed to maximise the dissemination of the organism: each spore has the potential to reproduce a complete entity. There are two main types of spores, which are classified according to their dispersal method as hydrophilic and hydrophobic. The former is dispersed by water and droplet splash and is therefore more active on rainy days, whereas the latter is transported by dry and hot air. Most common indoor fungi do not have active dispersion mechanisms, and the spores are mainly transferred through air movement caused by ventilation, occupants' activities or other disturbances.

Fungi are heterotrophic. Their metabolism relies only on the decay of organic materials to absorb nutrients. The digestion process usually occurs via an organic catalyst, meaning that fungi produce special enzymes that can transform the complex organic compound into simple digestible sugars. Fungi can metabolise a wide variety of simple and complex carbohydrates, such as cellulose, starch and lignin, making building materials a nutrient-rich environment suitable for growth. Due to the high number of spores released by each organism, mutations are common and Fungi can quickly adapt to harsh environments (Torvinen et al., 2006) and mutate,

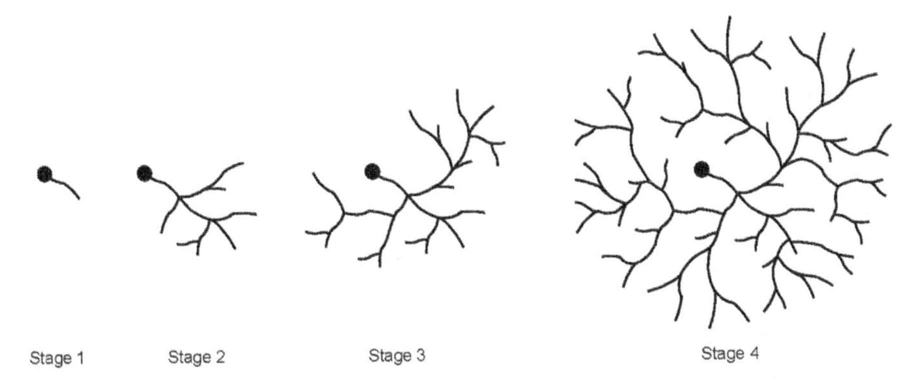

Figure 4.1 Diagrammatic representations of the fungal growth process. Stage one: germ tube initiates the growth process; stage 2: germ tube germinates; stage 3 and 4: mycelium is formed.

producing new enzymes to metabolise complex materials (Vacher, Hernandez, Bärtschi, & Poussereau, 2010). This characteristic makes it very difficult to predict mould growth, highlighting the importance of considering moisture-prevention strategies from the early design stage. However, indoor mould germination, growth and production of spores are highly dependent on a combination of factors, such as environmental conditions, climate and substrates. Therefore the species isolated in different contexts may vary.

The taxonomy of fungi is based on binomial nomenclature that indicates both the genus and specific strain. The most common mould species isolated in indoor environments are the genera *Penicillium*, *Aspergillus*, *Chaetomium*, *Stachybotrys* and *Cladosporium* (Grant, Hunter, Flannigan, & Bravery, 1989; Hunter, Grant, Flannigan, & Bravery, 1988; Hyvärinen, Meklin, Vepsäläinen, & Nevalainen, 2002) and the strains *Alternaria alternata* and *Ulocladium chartarum* (Gutarowska & Czyżowska, 2009). The literature identifies a wide variety of fungi that have been found in indoor environments (mainly residential buildings) around the world. An indicative list is presented in Table 4.1.

Studies show that *P. chrysogenum* is found equally in cold and warm environments, whereas among the *Aspergillus*, the most common is *A. versicolor*. All hygroscopic materials can be affected by mould growth. *Cladosporium* spp. is the most common on external building facades (Shirakawa, John, Gaylarde, Gaylarde, & Gambale, 2004; Shirakawa, Gaylarde, Gaylarde, John, & Gambale, 2002), whereas *Aureobasidium pullulans* (Eaton & Hale, 1993; Gobakken & Westin, 2008; Sandberg, 2008) usually infests exposed outdoor wood. Timber-based indoor materials are highly sensitive to mould growth, especially *A. versicolor* (Hyvärinen et al., 2002), *A. alternata*, *Cladosporium* spp., *Penicillium* spp. (Khan & Karuppayil, 2012), *Coniophora puteana* (also known as wet rot) and *Serpula lacrymans* (dry rot) (Viitanen, 1994). These fungi can degrade the material and compromise its structural stability. Concrete-based materials are mostly affected by *Aspergillus* spp., in particular, *A. melleus*, *A. niger* and *A. ochraceus* (Hyvärinen

Table 4.1 Indoor fungi species isolated from building materials and indoor environments.

Location	Dominant species belonging to *Penicillium*, *Aspergillus*, *Chaetomium*, *Stachybotrys* and *Cladosporium*	Additional isolated fungi	Reference
Subartic	*A. ochraceus*, *A.* glaucus	*Phoma violacea*, *Ulocladium atrum*	Lie, S. K., Thiis, T. K., Vestøl, G. I., Høibø, O., & Gobakken, L R. (2019). Can existing mould growth models be used to predict mould growth on wooden claddings exposed to transient wetting?. Building and Environment, 152, 192–203.
			Salonen, H., Lappalainen, S., Lindroos, O., Harju, R., & Reijula, K. (2007). Fungi and bacteria in mould-damaged and non-damaged office environments in a subarctic climate. Atmospheric Environment, 41(32), 6797–6807.
Europe—cold climate	*A. ustus*, *A. versicolor*, *C. sphaerospermum*, *C. ghbosum*, *P. chrysogenum*, *S. chartarum*	*Trichoderma harzianum*, *Chaetomium globosum*	Nielsen, K. F, Holm, G., Uttrup, L P., & Nielsen, P. A. (2004). Mould growth on building materials under low water activities. Influence of humidity and temperature on fungal growth and secondary metabolism. International Biodeterioration & Biodegradation, 54(4), 325–336.
			Gutarowska, B., Sulyok, M., & Krska, R. (2010). A study of the toxicity of moulds isolated from dwellings. Indoor and Built Environment, 19(6), 668–675.
			Wady, L., Bunte, A., Pehrson, C, & Larsson, L. (2003). Use of gas chromatography-mass spectrometry/solid phase microextraction for the identification of MVOCs from moldy building materials, Journal of microbiological methods, *52*(3), 325–332.

			Gravesen, S., Nielsen, P. A., Iversen, R., & Nielsen, K. F. (1999). Microfungal contamination of damp buildings--examples of risk constructions and risk materials. Environmental Health Perspectives, 107(Suppl. 3), 505–508.
			Nielsen, K. F., Thrane, U., Larsen, T. O., Nielsen, P. A., & Gravesen, S. (1998). Production of mycotoxins on artificially inoculated building materials. International biodeterioration & biodegradation, 42(1), 9–16.
Europe—temperate climate	*A. versicolor*, *A. fumigatus*, *A. niger*, *C. sphaerospermum*, *P. chrysogenum*, *P. olsonii*, *P. expansum*, *S. chartarum*	*Aurebadisium pullulans*, *Rhizopus* spp., *Trichoderma atroviride*	Reboux, G., Bellanger, A. P., Roussel, S., Grenouillet, F., Sornin, S., Piarroux, R., . . . & Millon, L. (2009). Indoor mold concentration in Eastern France, Indoor Air, 19(6) 446–453.
			Wouters, I. M., Douwes, J., Doekes, G., Thorne, P. S., Brunekreef, B., &. Heederik, D, J. (2000). Increased levels of markers of microbial exposure in homes with indoor storage of organic household waste. Applied and Environmental Microbiology, 66 (2), 627–631.
			Haas, D., Habib, J., Galler, H., Buzina, W., Schlacher, R., Marth, E., & Reinthaler, F. F. (2007). Assessment of indoor air in Austrian apartments with and without visible mold growth. Atmospheric Environment, 41(25), 5192–5201.
			Woodcock, A. A., Steel, N., Moore, C. B., Howard, S. J., Custovic, A., & Denning, D. W. (2006). Fungal contamination of bedding. Allergy, 61(1), 140–142.

(Continued)

Table 4.1 (Continued)

			de Ana, S. G., Torres-Rodríguez, J. M., Ramirez, E. A., Garcia, S. M., & Belmonte-Soler, J. (2006). Seasonal distribution of Alternaria, Aspergillus, Cladosporium and Penicillium species isolated in homes of fungal allergic patients. Journal of investigational allergology & clinical immunology, 16(6), 357–363.
			Benndorf, D., Müller, A., Bock, K., Manuwald, O., Herbarth, O., & Von Bergen, M. (2008). Identification of spore allergens from the indoor mould Aspergillus versicolor. Allergy, 63(4), 454–460.
			Polizzi, V., Adams, A., Picco, A. M., Adriaens, E., Lenoir, J., Van Peteghem, C.,. . . & De Kimpe, N. (2011). Influence of environmental conditions on production of volatiles by Trichoderma atroviride in relation with the sick building syndrome. Building and Environment, 46(4), 945–954.
Australia	*A. niger*, *C. cladosporioides*	*Fusarium* spp., *Botrytis* spp., *Rhizopus* spp., *Epicoccum* sp	Kanaani, H., Hargreaves, M., Ristovski, Z., & Morawska, L. (2009). Fungal spore fragmentation as a function of airflow rates and fungal generation methods. Atmospheric Environment, 43(24), 3725–3735.
			Chakraborty, S., Sen, S. K., & Bhattacharya, K. (2000). Indoor and outdoor aeromycological survey in Burdwan, West Bengal, India. Aerobiologia, 16(2), 211–219.
Japan	No particular dominant specie isolated	*Aurebadisium* spp., *Eurotium* spp., *Rhodotorula* spp.	Takigawa, T., Wang, B. L., Sakano, N., Wang, D. H., Ogino, K., & Kishi, R. (2009). A longitudinal study of environmental risk factors for subjective symptoms associated with sick building syndrome in new dwellings. Science of the total environment, 407 (19), 5223–5228.

India	No particular dominant specie isolated	*Rhizopus* spp., *Curvularia* spp., *Fusarium* spp., *Drechslera* spp., *Helminthosporium solani*	Chakraborty, S., Sen, S. K, & Bhattacharya, K. (2000). Indoor and outdoor aeromycological survey in Burdwan, West Bengal, India. Aerobiologia, 16(2), 211–219.
Middle East	*A. flavus*, *A. fumigatus*, *A. penicilloides*, *A. repens*, *P. glabrum*	*Acremonium* spp., *Botryodiplodia* spp., *Circenells* spp., *Myrothecium* spp., *Syncepalastrum* spp., *Cercospora*, *Drechslera* spp., *Embellisia*, *Fusarium* spp., *Rhizopus* spp., *Torula*, *Scytalidium* spp., *Curvularia* spp	Alwakeel, S. S., & Nasser, L. A. (2011). Indoor terrestrial fungi in household dust samples in Riyadh, Saudi Arabia. Microbiology Journal, 1(1), 17–24.
			Hasnain, S. M., Akhter, T., & Waqar, M. A. (2012). Airborne and allergenic fungal spores of the Karachi environment and their correlation with meteorological factors. Journal of Environmental Monitoring, 14(3), 1006–1013.
			Bokhary, H. A., & Parvez, S. (1995). Fungi inhabiting household environments in Riyadh, Saudi Arabia. Mycopathologia, I30(2), 79–87.
North America	*P. chrysogenum*, *P. crustosum*, *P. aurantiogriseum*	Bipolaris	Morey, P. R., Hull, M. C, & Andrew, M. (2003). El Nino water leaks identify rooms with concealed mould growth and degraded indoor air quality. International Biodeterioration & Biodegradation, 52 (3), 197–202.

(*Continued*)

Table 4.1 (Continued)

			Trout, D., Bernstein, J., Martinez, K., Biagini, R., & Wallingford, K. (2001). Bioaerosol lung damage in a worker with repeated exposure to fungi in a water-damaged building. Environmental Health Perspectives, 109(6), 641–644.
			Khan, Z. U., Khan, M. A. Y., Chandy, R., & Sharma, P. N. (1999). Aspergillus and other moulds in the air of Kuwait. Mycopathologia, 146(1), 25–32.
Uganda	No particular dominant specie isolated	*Mycosphoerella* yeasts, *Fusarium* spp., *Cochliobolus* spp.	Ismail, M. A., Chebon, S. K., & Nakamya, R. (1999). Preliminary surveys of outdoor and indoor aeromycobiota in Uganda. Mycopathologia, 148(1), 41–51.

From Brambilla, A., & Sangiorgio, A. (c. 2020). Mould growth in energy efficient buildings: Causes, health implications and strategies to mitigate the risk. *Renewable and Sustainable Energy Reviews*, 132. https://doi.org/10.1016/j.rser.2020.110093.

et al., 2002). In contrast, common plaster, gypsum board and wallpaper are sensitive to the attack of a broader range of fungi, including *Stachybotrys* spp., *Cladosporium* spp., *Ulocladium* spp. and *Aspergillus* spp. (Gravesen, Nielsen, Iversen, & Nielsen, 1999; Norge, 2014).

As the figure indicates, a wide range of different moulds can coexist, raising questions about the potential effects on humans. Indeed, some moulds that grow on building materials produce semi-volatile mycotoxin (Jarvis & Miller, 2005; Miller & McMullin, 2014), with potentially serious impacts on health.

4.1.2 Mites

Mites are microorganisms that live and proliferate in indoor environments. Their primary source of nutrition is human skin scale and water absorbed directly from moist air. House mites are adapted to live in relatively dry environments and can survive when relative humidity (RH) is as low as 50% (Arlian & Morgan, 2003). These features allow for rapid proliferation of house mites when favourable humidity conditions are met, considering that their main nutrients are always in abundant supply indoors. The concentration of house mites is higher in humid regions, and mould can be considered an indication of mite infestation, due to the similar humidity requirements (Peat et al., 1998). The main issue associated with house mites is the exacerbation of asthma in allergic subjects and potential sensitisation to respiratory conditions following long periods of exposure.

4.1.3 Bacteria

Bacteria are microscopic single-celled organisms with a very simple structure characterised by the absence of a nucleus or complex membranes. Actinobacteria are filamentous and can produce a consistent mycelium with a branching growth pattern, just like fungi. For this reason, they are also called actinomycetes and, in the past, were considered as an intermediate organism between bacteria and fungi. Actinobacteria are frequently found in damp buildings (Andersson et al., 1997; Hyvärinen et al., 2002; Schäfer, Jäckel, & Kämpfer, 2010; Suihko et al., 2009), especially on mould-damaged building materials (Lorenz, Kroppenstedt, Trautmann, Stackebrandt, & Dill, 2003), indicating the strong correlation between the two microorganisms. In buildings, the most common species are *Nocardia*, *Streptomyces*, *Amycolatopsis*, *Saccharoplyspora* and *Pseudonocardia* (Schäfer et al., 2010). Some of these can generate serious infections, allergic reactions and hypersensitivity pneumonitis (Lacey & Crook, 1988; McNeil & Brown, 1994; Minder & Nicod, 2005), thereby increasing the health hazards associated with damp buildings. Actinobacteria proliferation responds to the same factors as mould: temperature, substrate pH and moisture. Their optimal growth temperature ranges between 25°C and 30°C, but they can adapt to higher temperatures (up to 60°C) with high levels of RH (Edwards, 1993). Mycobacteria occur on a wide variety of substrates, such as ceramic tiles, wood and mineral insulation, and can also adapt easily to alkaline situations (Torvinen et al., 2006), increasing their ability to survive unfavourable conditions.

However, actinobacteria require higher RH to proliferate than fungi, and they are not as visible as mould. Thus assessment criteria and design requirements are usually based on mould growth.

4.1.4 Algae

Damp buildings contain another type of coloniser that infests the envelope surface: algae. Usually, these organisms belong to the families of *Chlorophytes* (also known as green algae) and *Cyanophytes* (blue-green algae) (Gaylarde & Gaylarde, 2005). The availability of water and the presence of natural light are the two main factors favouring their growth (Barberousse, Ruot, Yepremian, & Boulon, 2007), thus they are usually abundant on outdoor facades, where they are the principal cause of coloured stains and deterioration of cladding. In contrast to fungi, algae prefer rough, porous mineral substrates to organic ones (John, 1988), mainly due to their high water absorption. Algae are transported by the wind as spores and, once deposited on a surface, are retained in cavities until the presence of water and light favours their growth. When rain hits the spores, the algae are activated and spread on the surface, resulting in the characteristic striped appearance. Although algae contribute to the deterioration of building facades, they are mainly outdoor organisms and are therefore less harmful to human health.

4.2 Mould growth and health risks

Mould growth on building materials has a huge economic impact due to its role in the process of biodeterioration, which may lead to the need for costly reparation and early retrofit. For example, fungi are considered one of the major causes of colour change in exposed wood (Feist & Hon, 1984; Ridout, 2013), whereas brown rot can produce structural and physical deterioration in timber-based building components (Viitanen, 1994), requiring extensive remediation work. This problem is not confined to old constructions since, under the right conditions, substantial growth can also be expected in new buildings within a few years of construction (Gobakken & Lebow, 2010). Even more important, however, is the significant impact of fungi on the lives of the occupants of infested buildings. Mould is associated with both physical and psychological health problems and can affect work productivity (Liddell & Guiney, 2015). General symptoms associated with the presence of fungi in the indoor environment range from fatigue, reduced ability to concentrate and persistent nausea to extreme cases of cognitive impairment (Gordon et al., 2004). From a physical point of view, the proliferation of mould and bacteria may cause hypersensitivities, infections and toxicoses. Furthermore, microbial volatile organic compounds, the secondary organic compounds produced by fungi as they grow, can exacerbate the symptoms independently from exposure to the primary fungal mass (Khan & Karuppayil, 2012).

Research has identified the role of fungi in the aggravation of irritative health disorders, although it is difficult to establish a unique correlation between a specific exposure and particular adverse health effects, as different environmental factors, such as tobacco and air pollutants, may interfere. Thus most epidemiologic studies are conducted on children in order to isolate the effects of mould from those of tobacco, which may cause similar symptoms.

4.2.1 Hypersensitivities and allergies

The most numerous and varied health effects caused by exposure to mould are associated with the hypersensitivity diseases. These diseases usually develop in two stages. First, there is progressive sensitisation to the allergens, which prepare the human body to recognise foreign molecules. This is followed by the development of symptoms. The initial exposure to mould causes sensitisation, whereas the subsequent contact with allergens causes the production of histamine, which can irritate the upper airways or the lungs, leading to fever-like symptoms and asthma respectively. The majority of the adverse health effects caused by fungi are due to occupational exposure. It has been estimated that at least 10% of the population is sensitised to mould and house mite allergens, with the majority of subjects presenting with asthma (Kanchongkittiphon, Mendell, Gaffin, Wang, & Phipatanakul, 2014). In particular, a positive association has been established between *A. alternata* and the exacerbation of asthma symptoms (Peat et al., 1998), which can reach severe levels in children (Franks & Galvin, 2010).

Subjects living in indoor environments that are highly contaminated by fungi and mycotoxin may develop acute or chronic hypersensitivity pneumonitis and allergic alveolitis (Franks & Galvin, 2010; Singh, 1993). The symptoms in these cases are due to the immune response of the human body in fighting the allergens. Similar effects are caused by organic dust toxic syndrome, a noninfectious disease resulting from the inhalation of heavy organic compounds such as hyphae mixed with bacteria. Exposure to indoor mould is also associated with allergic rhinitis, sinusitis (Gutarowska, Sulyok, & Krska, 2010) and mucous membrane irritations, characterised by rhinorrhea, nasal congestion, sore throat and irritation of nose and eyes (Lanier et al., 2010).

4.2.2 Infections

Infections are generally caused by microorganisms that invade the living tissues of vulnerable hosts. Indoor fungi may be responsible for severe respiratory infections in subjects with serious diseases that compromise their ability to resist infections. This is particularly important in those buildings where the occupants' immune systems are already impaired, such as hospitals, aged care facilities or the homes of immuno-suppressed subjects. *Fusarium* spp. and *Aspergillus* spp. are particularly aggressive in regard to respiratory infections, with the latter being responsible for allergic broncho-pulmonary aspergillosis, pulmonary aspergillosis and pulmonary aspergilloma (Gutarowska et al., 2010; Singh, 1993). In highly humid environments, the bacteria responsible for Legionnaires' disease or other opportunistic infections can develop; however, the damage to the lung is less severe than in infections caused by fungi.

4.2.3 Toxicoses

Under favourable conditions, some fungi can produce mycotoxins, nonvolatile secondary metabolites. Their production depends on the fungus species, environmental conditions and the chemical composition of the substrate. A particular fungus may or may not produce mycotoxin depending on the substrate material, which makes it very difficult to predict its presence. For example, the *Aspergillus flavus* is known to produce a potent carcinogen, aflatoxin B1, only on rice, peanuts and soybeans, but fortunately not on building materials. Only a few studies have identified the presence of mycotoxins on building materials with adverse health symptoms, such as cough, eye irritation, joint ache, headache, fatigue, skin rash and irritation of airways (Khan & Karuppayil, 2012). However, because the amount that needs to be inhaled is very high, it is unlikely to occur in normal indoor environments.

4.2.4 Other health risks

Buildings infested by mould are usually characterised by damp indoor environments. High levels of RH are known to aggravate rheumatic diseases. These diseases are due to chronic inflammation in muscles and joints, and their severity is influenced by environmental conditions, including fungi, mites and organic compounds (Breda, Nozzi, De Sanctis, & Chiarelli, 2010). As in the case of fungal infections, these symptoms are found only in predisposed subjects. Although it might be rare to develop rheumatic diseases from mould, it is important to consider this aspect when designing aged care facilities, where rheumatic patients may experience severe exacerbation of symptoms.

4.3 Favourable conditions for growth

Fungi are extremely resilient organisms that can live and grow in a wide variety of conditions. The moment at which it becomes possible to see some germination activity under the microscope is referred to as the onset of mould growth (Isaksson, Thelandersson, Ekstrand-Tobin, & Johansson, 2010). The exact onset of growth depends on a wide range of factors, including water availability, environmental conditions (such as temperature and RH) and the substrate characteristics. It is therefore impossible to define a single onset condition, and a critical range of values pertaining to different parameters must be considered. These parameters are listed as follows.

- Environmental conditions RH and temperature
- Water activity
- Presence of nutrients on the substrate

4.3.1 Water activity

Among all the parameters, water availability is one of the most important. Fungi can grow in a broad range of thermal conditions, but they require a high level of humidity to initiate germination. The parameter that is used to measure water availability is water activity a_w, defined as the vapour pressure on the substrate divided by that of pure water at the same temperature, thus a_w falls within the range 0–1. Water activity is also called *equilibrated RH*, as in the condition of equilibrium between the substrate and the environment, a_w corresponds to the ambient RH divided by 100. In some studies conducted in climate chambers, equilibrium is reached and thus the two parameters correspond. However, it is essential to analyse the onset of growth according to RH and water activity separately, as it is not possible to assume that this condition is reached in real indoor environments. In fact, local differences in the ventilation or surface temperature, such as in the presence of thermal bridges, may generate a local micro-climate with a_w much higher than the indoor RH. For this reason, water activity is considered a better indicator of mould growth than RH.

Each fungus has its own a_w, but the range is generally between 0.78 and 0.8 at 10°C and 0.9 at 5°C (Nielsen, Holm, Uttrup, & Nielsen, 2004). The amount of nutrients on the substrate and the ageing of the spores also have an impact on the water activity value range. Indeed, higher water activity may compensate for lack of nutrients or lower temperatures, allowing fungi to grow in nonoptimal conditions; older spores require higher water availability to germinate. The germination process also requires a water activity 0.02 higher than the a_w critical for growth. Due to the complex interrelationship of all these parameters, it is very hard to define an exact critical limit for mould growth. Instead, it is described as a critical range. In most research studies, the minimum critical a_w for each fungus has been examined based on agar substrate. This is mainly because agar is nutrient-rich and its chemical composition is strictly controllable and thus easily reproducible. Fig. 4.2 below shows the a_w required by different fungi at 12°C and 25°C (Grant et al., 1989), and the minimum value found by other studies (Nielsen & Thrane, 2001).

The graph represents the water activity required on a nutrient-rich substrate; thus the a_w required on building materials may be higher. Generally, *Aspergillus* spp., *Penicillium* spp. and *Eurotium herbariorum* can grow with relatively low water availability, whereas *Stachybotrys chartarum* requires at least 0.85 a_w.

Water activity is also used as a parameter to classify indoor moulds, based on the a_w found in the laboratory on agar substrate. Three groups have been identified (Grant et al., 1989):

1. Primary colonisers: can grow with very low water activity, such as below 0.8, including some species of *Penicillium* spp., *Aspergillus* spp. and *Eurotium chartarum*;

2. Secondary colonisers: require average a_w between 0.8 and 0.9, including *Cladosporium*, *Alternaria* spp. and *Ulocladium* spp.;
3. Tertiary colonisers: require high level of a_w (generally above 0.9), usually met only in the presence of water; this category includes bacteria, actinomycetes, *Stachybotrys chartarum*, *Chaetomium* spp., *Trichoderma* spp. and *A. pullulans*.

Tertiary colonisers are particularly associated with situations in which water is actually present, such as leakage and condensation, since relative air humidity is not enough to achieve favourable conditions for growth. Although condensation and mould growth are not necessarily linked to each other, the presence of one is commonly taken as an indicator of the other. *Stachybotrys chartarum* is usually associated with high water availability and requires a nutrient-rich substrate to grow. Because its spores are rarely airborne, it is very difficult to isolate it from air

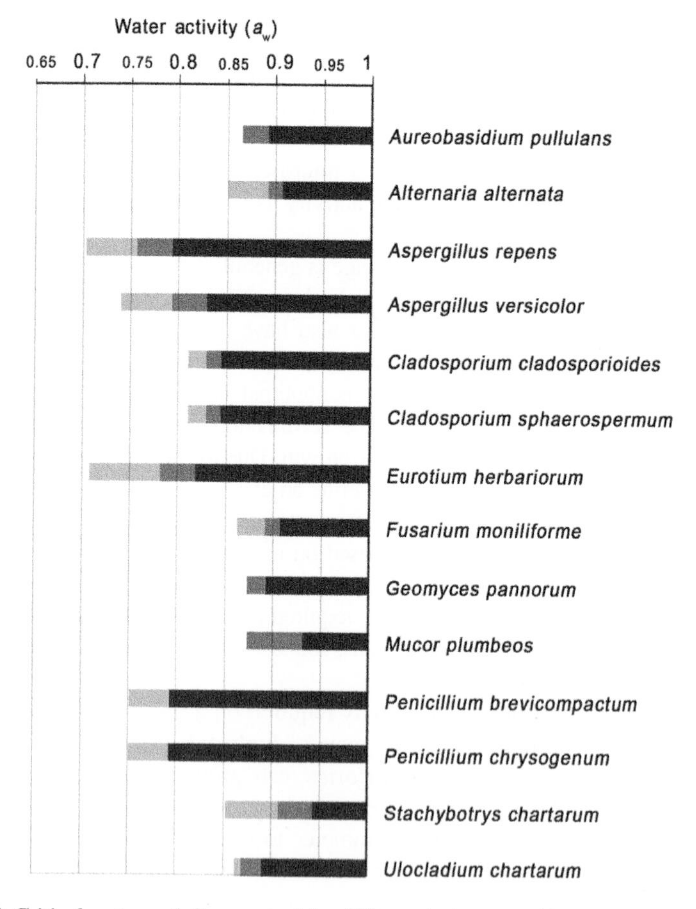

Figure 4.2 Critical water activity required by different indoor moulds to grow. Black colour water activity required at 12°C, dark grey colour: water activity required at 25°C, light grey colour: minimum water activity found in literature.

samples and a detailed investigation of the microbial community on a specific surface is required. Besides its high a_w, this fungus also requires long exposure to germinate, thus it is usually found in water-damaged buildings and on materials with high-cellulose content, such as drywall wallpaper, insulation backing, particleboards and fireboard. Another common tertiary coloniser, *A. pullulans*, is commonly found on continually damp surfaces, such as in bathrooms and kitchens.

Secondary and primary colonisers require less water to grow and are therefore more diffused indoors. *Alternaria* spp., generally classified as a secondary coloniser, is one of the most common indoor fungal strains. Its moderate requirements for water make it a very adaptable mould that can grow on most building materials, as well as carpets, wallpaper and synthetic materials. Another secondary coloniser, *Ulocladium* spp., is often isolated from painted surfaces, dust and air conditioners. Given the fact that fungal growth depends on several other factors, such as substrate and temperature, the classification may vary in different applications. Thus it is only to be considered as indicative of the moulds that could potentially grow in buildings.

4.3.2 Temperature

Spore germination is also influenced by temperature. Generally, lower temperatures are associated with slower growth rates. At the same time, however, lower surface temperatures are more likely to cause condensation, thereby increasing water availability. As such, temperature alone cannot be used to characterise mould growth.

For most fungi, the onset temperature of growth is between 5°C and 35°C (Selbmann et al., 2013); generally, indoor moulds require RH above 70% (Oliver, 1988) and a temperature range between 20°C and 30°C (S. Johansson, Wadsö, & Sandin, 2010; Lie, Vestøl, Høibø, & Gobakken, 2019), with an optimal mean value of 27°C (Johansson, Svensson, & Ekstrand-Tobin, 2013). However, fungal presence has also been observed in temperatures between 0°C and 50°C (Lie et al., 2019), and some *Penicillium* can grow on wood at −5°C (Land, Banhidi, & Albertsson, 1985). Fungi can germinate at a lower temperature when RH is high (Ayerst, 1969); therefore the most important factor to support growth is the combination of temperature and humidity, as one may compensate for the other.

4.3.3 Transient conditions

Environmental conditions in buildings are seldom stable, but fluctuate during the year due to changes in the external climate (rain, humidity and temperature) and the use of the building. This can result in periods of favourable and unfavourable conditions for mould growth. In particular, these fluctuating conditions have a negative impact on the development of fungi, delaying their growth (Gradeci, Labonnote, Time, & Köhler, 2017b). The frequency and sequence of favourable and unfavourable conditions can also influence mould growth, as does the duration of the fluctuations, as longer periods imply a faster growth rate (Viitanen & Bjurman, 1995).

Time of wetness (TOW) is a parameter that can be used to describe this phenomenon. It is defined as the ratio between the wet period (RH above 80%) and the total time period considered. Mould growth is observed to start for TOW equal to 0.17, and it increases significantly for TOW above 0.5. It is important to note that the variation in different parameters has a different potential effect on mould growth. For example, even small variations in RH have a greater impact on mould growth than significant variations in temperature (P. Johansson, Bok, & Ekstrand-Tobin, 2013).

4.3.4 Substrate

Besides water, humidity and temperature, the type of substrate also has a great influence on mould growth. The concentration of organic carbon in the substrate material, such as cellulose or lignin, is an indicator of favourable conditions for fungal growth. Although fungal diversity is greater in this type of material than in other types (Hyvärinen et al., 2002), under the right conditions, a wide variety of materials can support mould growth.

In buildings, timber-based materials, plasterboard and wallpaper are among the most susceptible due to their chemical composition, which offers highly digestible components. Their hygroscopic nature facilitates water retention and high humidity, creating favourable conditions for fungi to grow and germinate. However, the minimum RH content differs from one material to the other (Fig. 4.3), reflecting the availability of nutrients and the ease of digestion of the substrate itself.

Timber is naturally subject to mould attacks, but the risk and onset conditions vary greatly between different species or within different parts of the same tree trunk: for example, sapwood supports mould growth better than heartwood (Gobakken & Westin, 2008; Köse & Taylor, 2012; Theander, Bjurman, & Boutelje, 1993). This might be due to the higher presence of nutrients in the former or the toxic extractives naturally released by the latter.

Generally, growth is also influenced by the porosity of the wood species (Becker, Puterman, & Laks, 1986) but not by its density (Brischke, Bayerbach, & Otto Rapp, 2006). The humidity and nutrient conditions at the surface depend on its quality and treatment. Thus kiln-dried wood is more susceptible to mould growth than resawn timber, as it carries a large amount of nitrogen and carbohydrates on the surface, which mould can use as nutrients.

Paper-based coverings and paint also affect the capacity of a material to support growth, as paper contains high levels of organic carbon and water-based paints are highly susceptible to mould. Also, paints with low-vapour permeability can increase the level of water activity and the risk of mould growth. Materials that are usually paper-covered include plasterboard, probably the most widely used building component. Plasterboard is made of gypsum paste sandwiched between two layers of cardboard, cemented together by starch glue and gypsum crystals that penetrate the cardboard surface during the drying process. These cardboard sheets are highly susceptible to mould, and even relatively low humidity can trigger growth. This is important because plasterboard is highly hygroscopic and can absorb humidity from

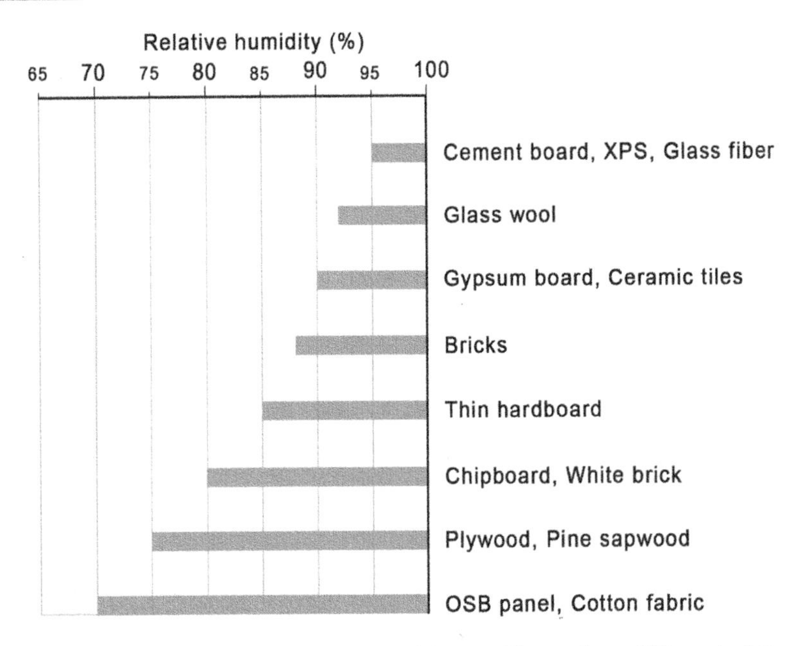

Figure 4.3 Minimum relative humidity required for mould growth on different building materials.

the air, with a consequent increase in the availability of water on the surface. The most common fungi found on the surface of plasterboards are *Stachybotrys chartarum*, *Penicillium* spp., *Aspergillus pp.*, *Chaetomium* spp. and *Ulocladium* spp. (Gravesen et al., 1999; Norge, 2014). The granularity and surface treatment of the material can also affect mould growth, as they can increase the retention of dust, which usually contains nutrient-rich human skin scales and significant concentrations of fungal spores. For this reason, some fungi have also been isolated on oil- or solvent-based paints, which can provide a good growing medium, especially for *A. fumigatus* and *A. versicolor*. In some cases, fungi can cause damage to the entire layers through eruptions, chips and cracks.

Under the right conditions, even materials such as aluminium or synthetic polymers can experience mould growth (M. A. Shirakawa et al., 2002; Vacher et al., 2010), whereas rubber and plastic are subject to biodeterioration from mould, which causes microcracking and loss of plasticity.

4.4 Mould growth assessment: models

The potential for mould infestation on a substrate depends on several interrelated factors. This interdependence means that mould growth is a highly complex living system that is very difficult to describe and predict. Due to this complexity, several models have been developed to assess the risk of mould during the design phase of a building. These models are based on different assumptions, use different methodologies, have different limitations and levels of accuracy, and classify the results in different ways. Consequently, the results of assessment of mould risk performed according to the different models may vary even if the inputs are the same, with sometimes significant discrepancies (Gradeci, Labonnote, Time, & Köhler, 2018; Vereecken & Roels, 2012). Therefore it is important to understand the various models' assumptions and assessment methods to determine their applicability and validity. This section discusses the most widely used and accepted models. Currently, only two of these—the VTT and biohygothermal models—are available as an easy-to-use add-on in the popular hygrothermal software (refer to Chapter 6: Hygrothermal modelling—*WUFI*) and are accordingly the best known assessment models.

4.4.1 VTT model

The VTT model (Hukka & Viitanen, 1999) is a mathematical model based on a regression analysis of empirical data obtained through an experimental programme. The first formulation of this model was only valid for and applicable to timber materials and could not be extended to other substrates. A revised version (Ojanen et al., 2010) addressed this shortcoming and extended the model to other building materials, based on results from new experiments using spruce board with glued edges, concrete, aerated concrete, cellular concrete, polyurethane with paper and polished surface, glass wool, polyester wool and expanded polystyrene.

The model was developed by assessing mould germination and growth through a visual inspection to identify the percentage of mould coverage on the substrate. This approach neglects the specific correlation of growth within the cell activities and relies only on subjective assessment.

The model correlates temperature and RH to the risk of mould growth, which depends on the sensitivity of the substrate. The risk is expressed by a mould index, which ranges between 0 and 6. Germination is associated with an index equal to 1, whereas the other values are associated with the percentage of mould coverage of the experimental samples, with 6 being 100%. The ability of this model to capture a nuance of different possibilities, or gravity, of mould occurrence has been used to update the standard ASHRAE 160, switching from a pass/fail to a risk-ranking approach.

Substrate materials are classified into the following four categories of sensitivity, based on generalisation from experimental findings.

1. Very sensitive: untreated wood or materials that offer optimal nutrients;
2. Sensitive: glued wooden boards, paper-coated products;
3. Medium resistant: cement and plastic-based materials, mineral fibres;
4. Resistant: glass and metal, materials with antifungal surface treatment.

Favourable environmental conditions for growth are defined by the critical value of relative humidity (RHcrit), which refers to the lowest humidity that allows mould to grow on a substrate after a certain exposure time. RHcrit is a function of temperature, and it assumes the lowest value at 20°C: this is 80% for material in classes 1 and 2 and 85% for the others. The final result of growth assessment performed over a certain period of time is given by the incremental value of the mould index assessed at hourly time steps. For each time step, the conditions are assessed against RHcrit, and if the favourable conditions for mould germination are not met, the model allows for either a delay or a decline in the mould index. The final value is then compared to the reference values (from 0 to 6) given by the model.

The VTT model is widely used to determine germination risk, but is not suitable for predicting growth rate. Although this model is based on empirical results, it presents some important limitations that need to be acknowledged in order to determine its applicability. The first shortcoming is the experimental sample size. Although the substrate is one of the main factors influencing mould growth, the model generalises the results for a limited number of materials. Additionally, no finishing material was tested except for untreated wood, which is common in northern Europe but less so in other parts of the world. Another limitation relates to the fungal species. Both the analysis of the samples and the classification matrix aim to describe the behaviour of mould as a generic entity, without taking into consideration the type of mould used in the study and its particular behaviour. The main limitation, however, relates to climate. The VTT model assumes that mould growth is not possible for RH below 80%, which is a significant over-simplification. Moreover, the model has not been verified for temperatures below 0°C and under realistic fluctuating conditions.

4.4.2 Biohygrothermal model

The biohygrothermal model correlates the growth phenomenon with the moisture balance at the spore level, considering the transient conditions of humidity and temperature at the boundary. It is based on a specific diagram, known as an isopleth system, which expresses favourable growth conditions in relation to temperature and RH. This model accounts for substrate type, climatic conditions and the minimum requirements for spore germination.

An isopleth is a curve that expresses the correlation between temperature, humidity and exposure time to determine mould growth. Each fungus is characterised by its own isopleth, which is determined by observation and experimental tests. The isopleth system has served as the basis for several growth models (Vereecken & Roels, 2012). Of these, the biohygrothermal model takes account of transient effects and the impact of drying on growth. This makes it the most realistic model.

The isopleths generate two different regions in the graph (Fig. 4.4), separating the zones with favourable and unfavourable growth conditions. Based on the graph, it is possible to determine if a specific combination of temperature and RH generates optimal conditions for mould to germinate and grow. The novelty of the biohygorthermal model lies in the definition of a second set of isopleths, used to express the growth rate. Thus the model uses the first isopleths to determine the germination conditions and the second to quantify the growth once germination is allowed.

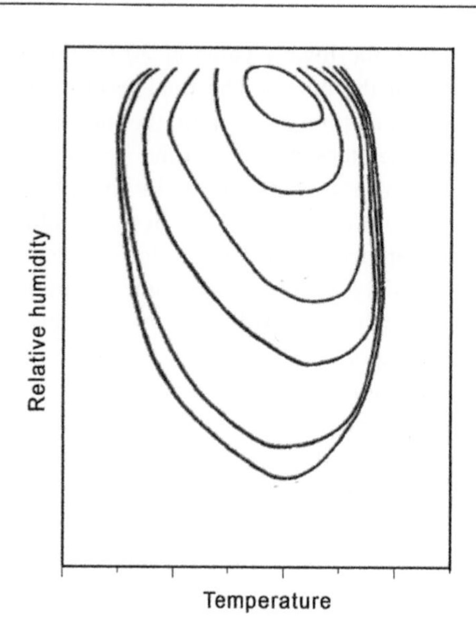

Figure 4.4 Example of isopleth–generic curve not referred specific fungus.

In this model, growth is allowed when the moisture content within a spore, which depends on RH of the air, reaches the critical value, which depends on temperature, RH and substrate sensitivity. This assessment is done based on the lowest isopleth for mould (LIM) curve, which was developed from experimental results. LIM does not characterise one specific fungus, but is representative of a specific cluster of fungal species, grouped based on their toxicity for human health. The LIM is generated as a composition of the isopleths of the fungi in each class, based on the lowest segment possible. Four classes are identified:

A. highly pathogenic—not allowed in buildings
B. pathogenic—long exposure may cause allergic reactions
C. not dangerous to health
K. additional class to represent the risks associated with *A. fumigatus, A. flavus*, and *Stachybotrys Chartarum*

For each class, representative strains have been chosen to describe the fungal behaviour as their isopleths are closely similar to the general LIM of their class. *Aspergillus versicolor* is representative for class A, whereas *A. amstelodami, A. candidus, A. ruber* and *Wallemia sebi* are representative for class B/C, which are further clustered together as their LIM curves are very similar.

To account for substrate sensitivity, the LIM curves have been ulteriorly defined for the following materials groups:

o. optimal biological culture;
i. biologically recyclable materials (wallpaper, gypsum covered with paper);
ii. mat with porous structure such as rendering, mineral material, species of wood;
iii. not biodegradable or nutrient-rich and considered to allow no mould growth.

Thus the model is based on a matrix of input that allows us to specify the mould class and the substrate type, making it suitable for a number of applications. In the biohygrothermal model, LIMs are used to determine whether the favourable conditions for germination and growth are met, and unlike the VTT model, this allows us to determine the growth rate, expressed as the formation of mycelium in mm/day. This second assessment is performed using a second isopleth, which correlates the favourable conditions to the possible growth rate, defined based on experimental results. The final growth rate is the sum of the rates defined for each time step, usually set as 1 hour. When the conditions are unfavourable, the growth rate is arrested, accounting for a possible delay in the process. However, the model does not consider an eventual set-back of the organic process, resulting in a continuous growth rate of hundreds of millimetres, which is highly unrealistic.

Although this approach can easily be used for a wide range of applications, the simplification embedded in the model should be carefully analysed. The isopleths, in fact, were developed only for specific fungal strains, and generalisation to all moulds may introduce a significant source of error. Furthermore, the model requires as input the initial water content of the spore, which contributes to determine the onset time for germination. As regression of the growth is not considered, higher initial values would allow for faster development of the mycelium, impacting the final result. The biohygrothermal model's validity depends on the definition of the LIM curves, which are defined for RH from 0% to 100%, and in the temperature range 0°C–30°C. Therefore its applicability for cases outside this range must be carefully considered.

4.4.3 Mould resistance design model

The Mould Resistance Design (MRD) model (Isaksson et al., 2010) is based on dose-response logic and is used to evaluate mould germination. The model was developed from the results of an experimental programme with different moulds inoculated on several samples of wooden materials.

The model assumes that a dose (indicator) is a function of the average RH and temperature over a 12-hour period. Mould growth is allowed when the dose value is equal to or greater than the critical dose, defined as the time needed for a specific fungus to germinate when exposed to a reference climate, meaning a combination of constant temperature and RH. In terms of the ratio between a dose and the critical dose, a dose can assume values greater than 1 when favourable conditions are met or less than 1 if the conditions are unfavourable. Dry or cool periods are considered as a negative dose, accounting for both delay and set-back of growth.

One of the major limitations of the MRD model is related to the validation process, which was performed using the same set of experimental results that were used to develop the VTT model, thus embedding the same assumptions, limitations and errors. This model also fails to account for the influence of solar radiation or wind-driven rain and is therefore only valid for sheltered conditions, such as crawl spaces or attics. Additionally, it was developed for wooden materials, making its applicability quite limited. In regard to climatic considerations, the MRD model also presents some shortcomings and limitations. It does not account for situations where the RH is below 75% or the temperature above 30°C, thereby overlooking the possibility of mould germination under these conditions. Finally, the MRD model was developed to assess germination and is therefore not intended to be used as a method to evaluate mould growth.

4.4.4 Johansson's indices

Johansson's indices (Johansson et al., 2010) are based on a set of experimental data on mould growth on exposed rendered facades, characterised by different colours, orientations and thermal inertia values. The experiments took place in Sweden over 20 consecutive months. Based on the results, three different indicators of increasing complexity were developed to describe the correlation between mould growth and environmental conditions.

The first index only accounts for RH, which is compared with a critical value (set at 80%). Growth is allowed when this value is exceeded. This index assumes a constant temperature, suggested to be the indoor temperature, and it is therefore not meaningful for applications where the temperature fluctuates or for outdoor conditions. The second index is based on two different equations that account for the interaction between RH and temperature. The correlation was developed based on the available literature on the behaviour of the *Cladosporium* spp. The model was further developed in the third index, which includes a delaying function. A potential set-back of the mould is not considered. The third index is supposed to be more accurate, as it completes the other two with an increasing level of complexity. However, this model is based on the behaviour of one specific fungus, making its application quite limited, and it does not allow for growth at RH below 80%. Additionally, the only substrates considered are wooden materials that are largely used in Northern Europe.

4.4.5 Mould Germination Graph model

The Mould Germination Graph (MGG) model (Moon, 2005) expands the isopleths system, accounting for the history of mould exposure to certain environmental conditions. It links each isopleth with a required time of exposure to the conditions favourable for initiating the germination process, and it considers temperature and RH at the previous time step to be equally important in determining growth onset. Growth is allowed when the cumulated time under favourable conditions for growth exceeds the critical exposure time of the isopleth. Thus the final index can be expressed as the sum of risky days. As a result of this definition, the model can account for fluctuations in environmental conditions and possible delay in growth. However, a regression in growth is not allowed, as the final index is defined as a sum.

One issue in the MGG model is related to the time step used, as it is based on daily averages. Smaller time steps, such as hourly based, are usually preferred and considered to more accurately describe the transitory nature of the mould growth phenomenon. Furthermore, once the favourable conditions cease to be met, the growth is stopped immediately, thus failing to consider a possible cool-off period, with resulting underestimation. This is partially compensated by the fact that growth is assumed to return to full regime as soon as the conditions become favourable, although in this case the growth risk may be overestimated. The MGG model does not apply to temperatures below 0°C.

4.4.6 General limitations and probabilistic models

Numerous models are available to assess the risk of mould growth, some of which are very advanced and well established in the literature. However, given the significant number of buildings infested by mould around the world, it is vital that these models be updated and improved in order to develop an accurate prevention assessment tool to reduce the health hazards and associated remediation costs. All the available models have similar limitations, as they have been developed and validated on similar input data while applying different hypotheses or calculation methods.

One of the most important issues is related to climatic validity. Most of the models have been developed with a heating-dominated cool climate in mind, thus neglecting the effects of high temperatures on the growth process. Indeed, temperatures above 35°C or below 0°C have not been investigated, raising questions about the suitability of these models for hot and humid climates. Additionally, the experiments performed do not account for RH ranging from 70% to 80% under fluctuating conditions, which might represent a more realistic scenario. Accordingly, there is insufficient input data to extend and validate the models for a wider context (Gradeci et al., 2017b). One major limitation is that most of the models are based on experimental data collected in climate chambers using optimal substrate (agar) or untreated wood; very few studies have been based on typical materials used in building construction. To date, no laboratory test results have been considered reliable in predicting outdoor conditions (Lie et al., 2019). The models are usually intended for the application to sheltered or interior conditions and require only data on RH and temperature, overlooking the importance of transient wetting conditions such as rain penetration or condensation (Lie et al., 2019). However, microclimatic conditions on the surface under investigation may have a significant impact, leading to wide variability in the conditions over time and space.

All the models embed simplification and uncertainties related to the definition of the biological phenomenon, climate, exposure and the materials considered. To account for these, the development of new generation models is based on a probabilistic rather than a deterministic approach. For example, the Gobakken et al. model (Gobakken & Lebow, 2010; Gobakken, Høibø, & Solheim, 2010) expresses the risk of growth as a cumulative probability, based on temperature (above 5°C), RH (above 80%), wooden substrate, coating typology and exposure time. On the other hand, Gradeci et al. (Gradeci, Labonnote, Time, & Köhler, 2017a) propose a probabilistic method that combines the three major mould growth models (VTT, biohygrothermal and MRD) to take advantage of the strengths of each and compensate for their weaknesses. Furthermore, the lack of a standardised experimental framework makes it very difficult to generalise and compare the results obtained so far during the development of the different models. There is a need for additional laboratory tests, with broader boundary conditions and a wider spectrum of building materials. Standardised conditions would allow for comparison of the results of different tests undertaken by different studies to create a unified model.

A further complication is related to the use of mould growth models to predict the risk of biodeterioration in buildings during the design stage. Indeed, the models require a specific climate as input, defined at hourly steps. Usually, hygrothermal simulations use 1-year historical weather data, called the Moisture Design Reference Year (MDRY). However, 1 year is not a long exposure time, especially when compared to a building's service life. Further, even when the test is repeated multiple times, the MDRY is not able to capture the actual variability of climatic conditions. Therefore these types of climatic files are not able to provide reliable and realistic results, especially because the majority of the growth models does not allow for mould regression. To overcome this issue, the Gradeci et al. model also incorporates a stochastic representation of the exposure climate, which considers a wider distribution of probable climatic conditions.

However, the shift from a deterministic to a probabilistic approach also increases the level of complexity of the method itself, making the assessment difficult to apply at the early stage of the design process, when it is most needed.

4.5 Mould-free constructions: design for safety and remediation strategies

The presence of microorganisms in indoor environments is a natural phenomenon. It is impossible to create a spore-free environment, and there is no suggestion that architects should seek to design for it. The most successful strategy for a mould-free construction involves:

1. adopting a mould risk-conscious design approach, reducing the risks of leakage or condensation via appropriate design of the building elements, material and finishes and/or using mechanical devices to reduce the water content of the internal spaces;
2. changing occupants' behaviour to reduce the risk of mould growth through good management of the indoor environment.

4.5.1 Mould risk-conscious design

Mould growth is supported by the presence of two major factors: water and nutrients. Thus a mould risk-conscious design should aim to manage these two factors in ways that reduce the potential for combinations of conditions that favour spore germination and fungal growth.

The presence of water on building components depends on air humidity and leakage. The design should aim to minimise the risks of water leaking from the external space through the careful provision of water barriers and, at the same time, minimise the potential for surface and interstitial condensation, reduce thermal bridges and assess the risks of condensation and the amount of condensate inside each building element. In other words, the resistance of the envelope should be determined based on the principles of the hygrothermal process, accounting for moisture transfer.

An important step in the design that has an impact on both water and nutrients is materials selection. The materials selection process is driven by multiple factors, such as architectural intent, energy efficiency, durability and cost, but the risk of mould growth is not taken into account. In fact, new sustainability trends recommend the use of bio-based or natural materials, which are an optimal medium for mould growth, while a decision based on economic factors alone would likely suggest the extensive use of wooden chipboard or particleboard. In buildings, the main materials-related concern is where they are used, rather than the nature of the material itself. In the wrong location, they can support the rapid development of mould infestation, even a few years after construction. It is essential to use the right material in the right location to prevent fungal attack and early biodeterioration. Places that experience cyclical and recurrent levels of high humidity, such as wet-rooms, should be finished with low-hygroscopic cladding to reduce water uptake by the material and water availability at the surface. The same principle applies to all the wall components in highly sensitive locations. An example is the use of organic insulation panels as fire retardants in the proximity of ventilation systems, which can cause serious damage apart from the deterioration of the material itself. The spores can infiltrate the ventilation system and spread from one location to another, resulting in a highly infested indoor environment (Aisner, Schimpff, Bennett, Young, & Wiernik, 1976).

Increased ventilation can also help to reduce the spore concentration and relative indoor humidity, minimising growth potential. The use of a hygrothermal-based ventilation system can facilitate the correct ventilation rate, overcoming the possible negative effects of the incorrect use of the ventilation system by the building occupants. Ideally, the ventilation system should be designed not only to discharge the thermal load and meet health and hygiene requirements, but also address humidity. If correctly designed, the ventilation system should allow air exchange with the exterior only when there is lower humidity outdoors than indoors. Strengthening the ventilation system to discharge the humidity accumulated indoors is one of the most efficient measures to prevent fungal growth.

4.5.2 Management of the indoor environment

The correct management of the indoor environment is essential to minimise the risk of mould growth. Fungi can find favourable conditions for germination in highly humid environments and in the presence of nutrients, such as dust and human skin scales. In the cycle of favourable and unfavourable conditions, the mould itself can become the nutrient for new mould. Thus an optimal mould-free approach to indoor management targets the reduction of these factors.

Regular cleaning helps to reduce the nutrients present in dirt and dust and the concentration of spores that enter from outside. Particular attention should be paid to carpets and mattresses, as they are also the primary means of diffusion of mites (Khan & Karuppayil, 2012).

Internal RH not exceeding 65% minimises the amount of water available to spores. Besides the use of moisture-based mechanical ventilation systems, humidity control can be achieved in two ways: by reducing humidity through, for example, the installation of wet-zone fans, and by minimising moisture generation. Normal residential activities can increase indoor moisture and the potential for mould growth. Cooking, bathing, large numbers of indoor plants that require frequent watering, pet urine and clothes dryers vented to the indoors significantly increase indoor humidity. Awareness of the impact of human activities on mould growth can assist in reducing the risk. In this context, British regulations and the Australian handbook, Condensation in Buildings (Australian Building Code Board, 2019), provide guidelines for building users to minimise the risks and a series of provisions for builders to reduce the risks of moisture becoming trapped in the construction materials (refer to Chapter 7: Building codes and standards).

4.5.3 Remediation strategies

Prevention of mould growth is the only effective strategy to avoid biodeterioration. However, when fungi are already colonising an indoor environment, it is important to adopt the correct remediation measures.

It is widely believed that chemical cleaning, such as toxic wash with diluted bleach, can kill mould. However, surface cleaning is proven to be useless in eliminating fungal growth when the infestation is already advanced. Indeed, in a recent investigation, 10 different chemicals were used to remove mould from building materials, but none were effective in eradicating the infestation or destroying the mycotoxins produced by the infestation (Peitzsch, Bloom, Haase, Must, & Larsson, 2012). This is because fungi can penetrate degradable materials and produce resistance to surface treatments. The only strategy that can be adopted to totally eradicate mould infestation is an extensive environmental remediation and replacement of the compromised materials (Haverinen-Shaughnessy, Hyvärinen, Putus, & Nevalainen, 2008). In order to avoid a recurrence, an analysis must be undertaken to identify the source of the problem, such as leakage, condensation or low surface temperature due to thermal bridges. Depending on the outcome of this assessment, different strategies might be considered.

As part of the remediation strategy, in order to avoid the regrowth of mould, a viable solution can be the use of self cleaning applied coatings, that provides a toxic environment for the germination and growth of mould. These coatings, that usually employ titanium dioxide particles, can also improve the evaporation capacity of the water and the decomposition of dirt and organic contaminants, further reducing the risk.

In the case of mould growth due to surface or interstitial condensation, a common solution in cold climates is to add external insulation using External Thermal Insulation Composite Systems (ETICS). If installed correctly, the additional insulation reduces heat flow through the external walls and existing thermal bridges, resulting in an increased internal surface temperature and consequent reduction in the risk of condensation. This solution is particularly successful since it solves the problems of mould and degradation of materials as well as increasing the performance of the envelope in meeting more stringent energy targets.

4.6 Mycelium as innovative building material

Fungi can be highly dangerous for human health and can significantly reduce the service life of a building, resulting in considerable personal harm and economic loss. However, not all fungi are harmful for the built environment. On the contrary, some have shown great potential as construction materials. In recent years, increasing awareness of the negative impact of the construction sector on the environment has led to the development of innovative technologies for a sustainable future. One growing trend is to take advantage of the benefits of using organic or bio-based materials in construction.

Organic-based materials are materials that include an organic compound in their matrix. There are now mycelium-based building blocks and panels, which use the organic growth process of fungi to enhance the environmental properties of building materials. A major benefit is the low or negative embodied carbon associated with their life cycle. These materials are usually self-growing, requiring minimum input from humans and using nutrients from materials commonly available in nature. As such, they are generally self-generating, minimising the requirements for new resources. Certain organic-based materials can also grow *in loco*, reducing the heavy environmental cost of transportation and requiring a minimum amount of manufacturing. These multiple benefits extend from the production phase to the end of life. Compared to traditional building materials, they have lower carbon emissions during manufacture and can be composted after use, thus creating a natural recycling process in which the organic compound can become a growth matrix for the next generation of products. This characteristic is of particular interest because it follows the principle of the circular economy, one of the fundamental pillars of sustainability.

Such materials are completely revolutionising the way we think about buildings, which are now theoretically capable of growing naturally and adapting to our needs. However, they are subject to an element of stigma because of their usual association with unhealthy indoor environments and "cheap" construction. Nonetheless, their potential to reduce the environmental impact of buildings is of interest to researchers and architects around the world, who are investigating a viable way of introducing them to the market on a large scale.

4.6.1 Mycelium blocks as a regenerative architecture strategy

Mycelium mass is fast-growing and relatively easy to cultivate, as it needs only an organic material as nutrient and a combination of moisture and temperature that can easily be achieved at ambient conditions. Indoor moulds grow on building materials following a radial pattern, mainly bidimensional. However, when the spores are mixed into a matrix, mycelium can grow on the three dimensions, spatially connecting the hyphae and forming a sort of natural adhesive. When immersed in a nutrient-rich loose matrix, mycelium digests the organic compound and creates a compact material. Through this process, it is possible to create mycelium-based materials, such as building blocks or panels.

The properties of the final material depend on the combination of substrate and nutrient matrix, fungal strain and incubation. The fungal strains used for this application are not those responsible for indoor mould. The fungal species are selected based on the properties of their mycelium, which must be nontoxic and harmless to human health. The final product usually has good fire-retarding properties, is noncombustible and water resistant, making it highly suitable for construction. The hyphae created during growth are cross-connected to produce a redundant material that resembles a sponge. This characteristic means that is has two important beneficial properties for the construction industry: it makes the material extremely light and it traps air inside, improving thermal insulation. The low weight of mycelium-based material makes it very easy to handle, transport and assemble. The structural properties of this material make it unsuitable for load-bearing purposes. However, in comparison to light concrete blocks, it shows a greater compressive strength per volume. Indeed, the growing pattern, the fibrous structure, and the fibre interconnections give it good compressive strength, making it a perfect self-supporting material (Jones, Tien, Chaitali, Daver, & Sabu, 2017). The light weight and the spongy matrix also make it an interesting material in relation to hygrothermal and acoustic properties. Currently, the most viable use of mycelium-based material is for insulation, both thermal (Xing, Brewer, El-Gharabawy, Griffith, & Jones, 2018) and acoustic (Pelletier, Holt, Wanjura, Bayer, & McIntyre, 2013). Furthermore, its porous nature suggests it might be suitable as a moisture buffer to regulate indoor humidity levels and provide superior comfort.

The properties can also be customised to a certain extent by choosing the right combination of strain and substrate. The fungi most frequently used are *Ganoderma lucidum, Ganoderma tsugae, Ganoderma oregonense, Trametes versicolor, Piptoporus betulinus* and *Pleurotus ostreatus*. Of these, *P. ostreatus* and *G. lucidum* are most suitable for the purpose. The former is commonly known as oyster mushroom. It is edible and is one of the most suitable organisms for producing protein-rich food. Its rapid growth rate and broad adaptability to different substrates make it a perfect candidate as an "ingredient" for building materials. *G. Lucidum*, on the other hand, is preferred because it can grow on a wide variety of cellulose-based substrates, including those from waste. During digestion, it transforms the cellulose into chitin, the hard material in insect shells. Its ability to produce a mass that is hard on the exterior is highly valuable for construction purposes, as it makes the product resistant to impact (Fig. 4.5).

These two fungi can be grown in any shape, including as interlocking blocks or free forms, making it a very adaptable material for construction. The growing process is easy in principle, but it can be tricky. Although its basic requirements are a fertile substrate, a moist environment and partially controlled temperature, it is important to prevent any contamination during production to avoid infection by other moulds or bacteria (Fig. 4.6)

The initial matrix needs to be highly nutritious and is normally composed of agricultural waste, cardboard, woodchip, rice and wheat husks or sawdust; sometimes an inert fibrous aggregate is added to increase the final volume. Inoculation requires sterilised equipment, including the container in which the matrix is left to rest and the mixing tool, as well as gloves to protect against the transfer of possible contaminants. The inoculation period varies depending on the fungal strain, but the process is usually performed at 25°C and stable humidity. The slower the growth, the higher is the possibility of contamination, so a stable and protected environment, such as a climate or incubation chamber, is required. The mycelium can be left to grow indefinitely. However old mycelium is denser, which diminishes the insulation properties of the block. In order to stop the organic activity and the biological life cycle, the product needs to be cooked at a temperature between 70°C

Figure 4.5 Mycelium growing on the substrate to form the block.

Figure 4.6 Mycelium blocks after cultivation and curing. The dark sample shows infection from bacteria.

and 90°C. A 5-cm thick panel requires 1.5 hours; once the mycelium is cured, the material is ready for use.

4.6.2 Self-healing concrete: how bacteria can fix structural issues

Concrete is among the most commonly used building materials due to its structural properties, availability and adaptability. During its service life, concrete is subject to different stresses from both load and nonload factors, such as thermal stresses, shrinkage or a combination of factors. As a consequence, concrete may develop micro or nano-cracks. Although these cracks do not necessarily lead to significant change in structural performance, once formed, the crack can propagate and result in durability issues over time. An open crack provides a direct ingress for water and air into the concrete matrix, which can cause early degradation of the reinforcement due to chemical reactions that increase corrosion.

The techniques currently available to repair cracked concrete are expensive and extremely time-consuming. To respond to this challenge, researchers are investigating a new organic-based material capable of creating a self-healing concrete (Jonkers, 2011). This material takes advantage of the catalysis of certain bacteria, which transforms a nutrient into a product that can be used as an adhesive or binding aggregate in a compact matrix. Bacteria can transform calcium lactate into calcium carbonate which, in the specific case of concrete, can act as filler for the nano-cracks, sealing the material against air and water penetration. The self-healing concrete is made by mixing the matrix with encapsulated nutrition and bacteria, which can survive at a very low metabolic rate for up to 200 years (Schlegel & Zaborosch, 1993) and can then be reactivated by water and air. The bacteria used for this purpose are alkali-resistant organisms that can resist high mechanical and chemical stress for years, making them capable of surviving in the concrete. Self-healing concrete is fireproof and noncombustible and shows similar—if not superior—structural strength to normal concrete Jonkers, 2011; Jonkers & Schlangen, 2007; Vijay, Murmu, & Deo, 2017).

4.7 Future outlook

The increase in biodeterioration due to fungal growth in indoor environments is a matter of great concern around the globe. It has been estimated that the proportion of buildings affected by mould is as high as 45% in Europe, 40% in the United States, 30% in Canada and 50% in Australia. Although the conditions for mould germination and growth are known in principle, much needs to be done to eradicate this problem at its source.

Current growth models do not correctly describe the onset conditions, as they are all based on a limited amount of data that neglects issues of transient wetting and wind-driven rain. The lack of standardised conditions and assumptions in the models used in the mould testing process means that the results of various studies are often incomplete or are not applicable to the codes and simulation software used for mould assessment. Recent years have seen the development of software that is better able to simulate condensation and moisture in the built environment. However, the absence of reliable models for mould growth is one of the main reasons why the results of these analyses are still considered unreliable and open to interpretation. Further, the probabilistic approaches that have been suggested by research results are highly complex and difficult to apply during the design stage, as they require highly specialised expertise.

As well, global warming raises questions about their suitability for future climates, due to their temperature-related limitations. New research is needed to provide evidence for warmer and subtropical climates and to inform the development of building codes based on prevention-oriented design strategies.

Good design, however, is not sufficient on its own to prevent mould growth. It is also necessary to educate building occupants and construction companies about practices that increase indoor moisture and therefore create a favourable environment for mould proliferation.

References

Aisner, J., Schimpff, S. C., Bennett, J. E., Young, V. M., & Wiernik, P. H. (1976). Aspergillus infections in cancer patients: Association with fireproofing materials in a new hospital. *JAMA*, *235*(4), 411–412.

Allsopp, D., Seal, K. J., & Gaylarde, C. C. (2004). *Introduction to biodeterioration*. Cambridge University Press.

Andersson, M. A., Nikulin, M., Köljalg, U., Andersson, M., Rainey, F., Reijula, K., ... Salkinoja-Salonen, M. (1997). Bacteria, molds, and toxins in water-damaged building materials. *Applied and Environmental Microbiology, 63*(2), 387–393.

Arlian, L. G., & Morgan, M. S. (2003). Biology, ecology, and prevalence of dust mites. *Immunology and Allergy Clinics of North America, 23*(3), 443–468.

Australian Building Code Board. (2019). Handbook: Condensation in buildings.

Ayerst, G. (1969). The effects of moisture and temperature on growth and spore germination in some fungi. *Journal of Stored Products Research, 5*(2), 127–141.

Barberousse, H., Ruot, B., Yepremian, C., & Boulon, G. (2007). An assessment of façade coatings against colonisation by aerial algae and cyanobacteria. *Building and Environment, 42*(7), 2555–2561.

Becker, R., Puterman, M., & Laks, J. (1986). The effect of porosity of emulsion paints on mould growth. *Durability of Building Materials, 3*(4), 369–380.

Borrego, S., & Perdomo, I. (2012). Aerobiological investigations inside repositories of the National Archive of the Republic of Cuba. *Aerobiologia, 28*(3), 303–316.

Breda, L., Nozzi, M., De Sanctis, S., & Chiarelli, F. (2010). *Laboratory tests in the diagnosis and follow-up of pediatric rheumatic diseases: an update, Seminars in arthritis and rheumatism* (40, pp. 53–72). Elsevier.

Brischke, C., Bayerbach, R., & Otto Rapp, A. (2006). Decay-influencing factors: A basis for service life prediction of wood and wood-based products. *Wood Material Science and Engineering, 1*(3–4), 91–107.

Eaton, R. A., & Hale, M. D. (1993). *Wood: Decay, pests and protection.* Chapman and Hall Ltd.

Edwards, C. (1993). Isolation properties and potential applications of thermophilic actinomycetes. *Applied Biochemistry and Biotechnology, 42*(2–3), 161–179.

Feist, W. C., & Hon, D. N.-S. (1984). Chemistry of weathering and protection. *The Chemistry of Solid Wood, 207*, 401–451.

Franks, T. J., & Galvin, J. R. (2010). Hypersensitivity pneumonitis: Essential radiologic and pathologic findings. *Surgical Pathology Clinics, 3*(1), 187–198.

Gaylarde, C. C., & Gaylarde, P. M. (2005). A comparative study of the major microbial biomass of biofilms on exteriors of buildings in Europe and Latin America. *International Biodeterioration & Biodegradation, 55*(2), 131–139.

Gobakken, L. R., Høibø, O. A., & Solheim, H. (2010). Factors influencing surface mould growth on wooden claddings exposed outdoors. *Wood Material Science and Engineering, 5*(1), 1–12.

Gobakken, L. R., & Lebow, P. K. (2010). Modelling mould growth on coated modified and unmodified wood substrates exposed outdoors. *Wood Science and Technology, 44*(2), 315–333.

Gobakken, L. R., & Westin, M. (2008). Surface mould growth on five modified wood substrates coated with three different coating systems when exposed outdoors. *International Biodeterioration & Biodegradation, 62*(4), 397–402.

Gordon, W. A., Cantor, J. B., Johanning, E., Charatz, H. J., Ashman, T. A., Breeze, J. L., ... Abramowitz, S. (2004). Cognitive impairment associated with toxigenic fungal exposure: a replication and extension of previous findings. *Applied Neuropsychology, 11*(2), 65–74.

Gradeci, K., Labonnote, N., Time, B., & Köhler, J. (2017a). A probabilistic-based approach for predicting mould growth in timber building envelopes: Comparison of three mould models. *Energy Procedia, 132*, 393–398.

Gradeci, K., Labonnote, N., Time, B., & Köhler, J. (2017b). Mould growth criteria and design avoidance approaches in wood-based materials—a systematic review. *Construction and Building Materials, 150*, 77–88.

Gradeci, K., Labonnote, N., Time, B., & Köhler, J. (2018). A probabilistic-based methodology for predicting mould growth in façade constructions. *Building and Environment*, *128*, 33–45.

Grant, C., Hunter, C., Flannigan, B., & Bravery, A. (1989). The moisture requirements of moulds isolated from domestic dwellings. *International Biodeterioration*, *25*(4), 259–284.

Gravesen, S., Nielsen, P. A., Iversen, R., & Nielsen, K. F. (1999). Microfungal contamination of damp buildings—Examples of risk constructions and risk materials. *Environmental Health Perspectives*, *107*(3), 505–508.

Gutarowska, B., & Czyżowska, A. (2009). The ability of filamentous fungi to produce acids on indoor building materials. *Annals of Microbiology*, *59*(4), 807–813.

Gutarowska, B., & Piotrowska, M. (2007). Methods of mycological analysis in buildings. *Building and Environment*, *42*(4), 1843–1850.

Gutarowska, B., Sulyok, M., & Krska, R. (2010). A study of the toxicity of moulds isolated from dwellings. *Indoor and Built Environment*, *19*(6), 668–675.

Haverinen-Shaughnessy, U., Hyvärinen, A., Putus, T., & Nevalainen, A. (2008). Monitoring success of remediation: seven case studies of moisture and mold damaged buildings. *Science of the Total Environment*, *399*(1–3), 19–27.

Hukka, A., & Viitanen, H. (1999). A mathematical model of mould growth on wooden material. *Wood Science and Technology*, *33*(6), 475–485.

Hunter, C., Grant, C., Flannigan, B., & Bravery, A. (1988). Mould in buildings: The air spora of domestic dwellings. *International Biodeterioration*, *24*(2), 81–101.

Hyvärinen, A., Meklin, T., Vepsäläinen, A., & Nevalainen, A. (2002). Fungi and actinobacteria in moisture-damaged building materials—Concentrations and diversity. *International Biodeterioration & Biodegradation*, *49*(1), 27–37.

Isaksson, T., Thelandersson, S., Ekstrand-Tobin, A., & Johansson, P. (2010). Critical conditions for onset of mould growth under varying climate conditions. *Building and Environment*, *45*(7), 1712–1721.

Jarvis, B. B., & Miller, J. D. (2005). Mycotoxins as harmful indoor air contaminants. *Applied Microbiology and Biotechnology*, *66*(4), 367–372.

Johansson, P., Bok, G., & Ekstrand-Tobin, A. (2013). The effect of cyclic moisture and temperature on mould growth on wood compared to steady state conditions. *Building and Environment*, *65*, 178–184.

Johansson, P., Svensson, T., & Ekstrand-Tobin, A. (2013). Validation of critical moisture conditions for mould growth on building materials. *Building and Environment*, *62*, 201–209.

Johansson, S., Wadsö, L., & Sandin, K. (2010). Estimation of mould growth levels on rendered façades based on surface relative humidity and surface temperature measurements. *Building and Environment*, *45*(5), 1153–1160.

John, D. (1988). Algal growths on buildings: A general review and methods of treatment. *In Biodeterioration Abstracts*, *2*, 81–102.

Jones, M., Tien, H., Chaitali, D., Daver, F., & Sabu, J. (2017). Mycelium composites: A review of engineering characteristics and growth kinetics. *Journal of Bionanoscience*, *11*(4), 241–257. Available from https://doi.org/10.1166/jbns.2017.1440.

Jonkers, H. M. (2011). Bacteria-based self-healing concrete. *Heron*, *56*(1/2).

Jonkers, H.M., Schlangen, E. (2007). Self-healing of cracked concrete: A bacterial approach. Proceedings of FRACOS6: Fracture Mechanics of Concrete and Concrete Structures. Catania, Italy, 1821–1826.

Kanchongkittiphon, W., Mendell, M. J., Gaffin, J. M., Wang, G., & Phipatanakul, W. (2014). Indoor environmental exposures and exacerbation of asthma: An update to the

2000 review by the Institute of Medicine. *Environmental Health Perspectives, 123*(1), 6–20.

Khan, A. H., & Karuppayil, S. M. (2012). Fungal pollution of indoor environments and its management. *Saudi Journal of Biological Sciences, 19*(4), 405–426.

Klepeis, N. E., Nelson, W. C., Ott, W. R., Robinson, J. P., Tsang, A. M., Switzer, P., ... Engelmann, W. H. (2001). The National Human Activity Pattern Survey (NHAPS): a resource for assessing exposure to environmental pollutants. *Journal of Exposure Science and Environmental Epidemiology, 11*(3), 231.

Köse, C., & Taylor, A. M. (2012). Evaluation of mold and termite resistance of included sapwood in eastern redcedar. *Wood and Fiber Science, 44*(3), 319–324.

Krus, M., Sedlbauer, K., Zillig, W., Künzel, H. (2001). A new model for mould prediction and its application on a test roof. In II International Scientific Conference on The Current Problems of building physics in the rural building, Cracow, Poland.

Lacey, J., & Crook, B. (1988). Fungal and actinomycete spores as pollutants of the workplace and occupational allergens. *The Annals of Occupational Hygiene, 32*(4), 515–533.

Land, C., Banhidi, Z., Albertsson, A.C. (1985). Surface discoloring and blue staining by cold-tolerant filamentous fungi on outdoor softwood in Sweden. Material Und Organismen (Germany, FR).

Lanier, C., Richard, E., Heutte, N., Picquet, R., Bouchart, V., & Garon, D. (2010). Airborne molds and mycotoxins associated with handling of corn silage and oilseed cakes in agricultural environment. *Atmospheric Environment, 44*(16), 1980–1986.

Leung, M. H., & Lee, P. K. (2016). The roles of the outdoors and occupants in contributing to a potential pan-microbiome of the built environment: A review. *Microbiome, 4*(1), 21.

Liddell, C., & Guiney, C. (2015). Living in a cold and damp home: Frameworks for understanding impacts on mental well-being. *Public Health, 129*(3), 191–199.

Lie, S. K., Vestøl, G. I., Høibø, O., & Gobakken, L. R. (2019). Surface mould growth on wooden claddings—effects of transient wetting, relative humidity, temperature and material properties. *Wood Material Science & Engineering, 14*(3), 129–141.

Lorenz, W., Kroppenstedt, R., Trautmann, C., Stackebrandt, E., Dill, I. (2003). Actinomycetes in building materials. In International Conference Healthy Buildings, Singapore (pp. 583–589).

McNeil, M. M., & Brown, J. M. (1994). The medically important aerobic actinomycetes: Epidemiology and microbiology. *Clinical Microbiology Reviews, 7*(3), 357–417.

Miller, J. D., & McMullin, D. R. (2014). Fungal secondary metabolites as harmful indoor air contaminants: 10 years on. *Applied Microbiology and Biotechnology, 98*(24), 9953–9966.

Minder, S., & Nicod, L. (2005). *Exogen allergische Alveolitis (Hypersensitivitätspneumonitis). Swiss Medical Forum* (5, pp. 567–574). EMH Media.

Moon, H. J. (2005). *Assessing mold risks in buildings under uncertainty*. Georgia Institute of Technology.

Nielsen, K. F., Holm, G., Uttrup, L., & Nielsen, P. (2004). Mould growth on building materials under low water activities. Influence of humidity and temperature on fungal growth and secondary metabolism. *International Biodeterioration & Biodegradation, 54*(4), 325–336.

Nielsen, K.F., Thrane, U. (2001). Mould growth on building materials: Secondary matabolites, mycoxocins and biomarkers.

Norge, S. (2014). Paints and varnishes-Laboratory method for testing the efficacy of film preservatives in a coating against fungi (NS-EN 15457: 2014). Standard Norge, Oslo, Norway.

Ojanen, T., Viitanen, H., Peuhkuri, R., Lähdesmäki, K., Vinha, J., Salminen, K. (2010). Mold growth modeling of building structures using sensitivity classes of materials. Proceedings Buildings XI, Florida.

Oliver, A.C. (1988). Dampness in buildings.

Ortega-Calvo, J., Hernandez-Marine, M., & Sáiz-Jiménez, C. (1991). Biodeterioration of building materials by cyanobacteria and algae. *International Biodeterioration*, *28*(1–4), 165–185.

Peat, J. K., Dickerson, J., & Li, J. (1998). Effects of damp and mould in the home on respiratory health: A review of the literature. *Allergy*, *53*(2), 120–128.

Peitzsch, M., Bloom, E., Haase, R., Must, A., & Larsson, L. (2012). Remediation of mould damaged building materials—Efficiency of a broad spectrum of treatments. *Journal of Environmental Monitoring*, *14*(3), 908–915.

Pelletier, M., Holt, G., Wanjura, J., Bayer, E., & McIntyre, G. (2013). An evaluation study of mycelium based acoustic absorbers grown on agricultural by-product substrates. *Industrial Crops and Products*, *51*, 480–485.

Ridout, B. (2013). *Timber decay in buildings: the conservation approach to treatment.* Taylor & Francis.

Sandberg, K. (2008). Degradation of Norway spruce (Picea abies) heartwood and sapwood during 5.5 years' above-ground exposure. *Wood Material Science and Engineering*, *3* (3–4), 83–93.

Schäfer, J., Jäckel, U., & Kämpfer, P. (2010). Analysis of Actinobacteria from mould-colonized water damaged building material. *Systematic and Applied Microbiology*, *33* (5), 260–268.

Schlegel, H. G., & Zaborosch, C. (1993). *General microbiology*. Cambridge University Press.

Sedlbauer, K. (2001). *Prediction of mould fungus formation on the surface of and inside building components*. Fraunhofer Institute for Building Physics.

Sedlbauer, K. (2002). Prediction of mould growth by hygrothermal calculation. *Journal of Thermal Envelope and Building Science*, *25*(4), 321–336.

Selbmann, L., Egidi, E., Isola, D., Onofri, S., Zucconi, L., de Hoog, G. S., ... Balestrazzi, A. (2013). Biodiversity, evolution and adaptation of fungi in extreme environments. *Plant Biosystems—An International Journal Dealing with All Aspects of Plant Biology*, *147* (1), 237–246.

Shirakawa, M. A., Gaylarde, C. C., Gaylarde, P. M., John, V., & Gambale, W. (2002). Fungal colonization and succession on newly painted buildings and the effect of biocide. *FEMS Microbiology Ecology*, *39*(2), 165–173.

Shirakawa, M., John, V. M., Gaylarde, C., Gaylarde, P., & Gambale, W. (2004). Mould and phototroph growth on masonry facades after repainting. *Materials and Structures*, *37*(7), 472–479.

Singh, J. (1993). Biological contaminants in the built environment and their health implications: Biological contaminants in the built environment can raise concerns for the indoor air quality and the health of the building occupants in addition to the damage they can cause to the building structures, decorations and contents. *Building Research and Information*, *21*(4), 216–224.

Suihko, M.-L., Priha, O., Alakomi, H.-L., Thompson, P., Mälarstig, B., Stott, R., & Richardson, M. (2009). Detection and molecular characterization of filamentous actinobacteria and thermoactinomycetes present in water-damaged building materials. *Indoor Air*, *19*(3), 268–277.

Theander, O., Bjurman, J., & Boutelje, J. (1993). Increase in the content of low-molecular carbohydrates at lumber surfaces during drying and correlations with nitrogen content, yellowing and mould growth. *Wood Science and Technology*, *27*(5), 381–389.

Torvinen, E., Meklin, T., Torkko, P., Suomalainen, S., Reiman, M., Katila, M.-L., ... Nevalainen, A. (2006). Mycobacteria and fungi in moisture-damaged building materials. *Applied and Environmental Microbiology*, *72*(10), 6822–6824.

Vacher, S., Hernandez, C., Bärtschi, C., & Poussereau, N. (2010). Impact of paint and wall-paper on mould growth on plasterboards and aluminum. *Building and Environment*, *45* (4), 916–921.

Vereecken, E., & Roels, S. (2012). Review of mould prediction models and their influence on mould risk evaluation. *Building and Environment*, *51*, 296–310.

Viitanen, H. (1994). Factors affecting the development of biodeterioration in wooden constructions. *Materials and Structures*, *27*(8), 483–493.

Viitanen, H., & Bjurman, J. (1995). Mould growth on wood under fluctuating humidity conditions. *Material Und Organismen*, *29*(1), 27–46.

Viitanen, H., Vinha, J., Salminen, K., Ojanen, T., Peuhkuri, R., Paajanen, L., & Lähdesmäki, K. (2010). Moisture and bio-deterioration risk of building materials and structures. *Journal of Building Physics*, *33*(3), 201–224.

Vijay, K., Murmu, M., & Deo, S. V. (2017). Bacteria based self healing concrete—A review. *Construction and Building Materials*, *152*, 1008–1014.

WHO. (2009). WHO guidelines for indoor air quality: dampness and mould. World Health Organisation. https://www.euro.who.int/__data/assets/pdf_file/0017/43325/E92645.pdf?ua = 1

Xing, Y., Brewer, M., El-Gharabawy, H., Griffith, G., & Jones, P. (2018). *Growing and testing mycelium bricks as building insulation materials, IOP Conference Series: Earth and Environmental Science* (121, p. 022032). IOP Publishing.

Moisture buffering of building materials

5

Indoor humidity plays an important role in the assessment of occupant well-being and health. In addition to the health hazards associated with biodeterioration and mould growth on building materials due to damp, occupants' thermal perception depends significantly on the indoor humidity level. The human body's reaction to humidity influences the overall hygrothermal sensation. Sweating is a natural heat loss mechanism that relies on evaporation from the skin and which inspired the design of the evaporative cooling systems used in buildings. At higher ambient relative humidity, the air is closer to the dew point, which reduces evaporation through the skin, thus reducing capacity for natural cooling of the body. On the other hand, low relative humidity environments increase the body's transpiration and evaporation, resulting in dry skin and irritation. Based on these two thresholds, it is possible to identify an optimal relative humidity (RH) range for hygrothermal comfort of 30%–60% for air-conditioned buildings, where the indoor environment is fairly stable.

These interdependencies between thermal comfort and humidity are considered when the heating, ventilation and air conditioning (HVAC) system of a building is designed, and it is therefore important to account for all the moisture-related processes that occur in buildings by incorporating a system based on peak loads management. The ability of hygroscopic building materials to uptake and release moisture helps to buffer the indoor humidity and greatly influences the indoor environment by attenuating its variations, optimising occupants' comfort and reducing biological risks. This capacity helps to maintain the indoor quality at acceptable values and is essential in all applications that do not employ air conditioning systems. This is particularly important under unforeseen and extreme conditions, when the system is unable to completely de-humidify the environment. However, there is currently no agreed framework that encompasses current knowledge in the field.

This chapter discusses the interactions between the indoor environment and the building envelope, and considers their implications. It examines moisture buffering as a passive moisture management strategy by exploring relevant concepts, quantification methods and classifications. It investigates the phenomenon from both the micro-scale perspective of material properties and the macro-scale impacts on the whole building.

5.1 Moisture uptake and release

The interconnections between indoor moisture and the health of dwellers have been investigated since the early 1900s, when Ingersoll identified the first recognised

Moisture and Buildings. DOI: https://doi.org/10.1016/B978-0-12-821097-0.00002-3

case of moisture buffering (Ingersoll, 1913). By allowing the same quantity of water to evaporate each day in his own house, he discovered that cold days produced severe condensation effects, especially in upholstered furniture in the bedroom (Ingersoll, 1913; Svennberg, Lengsfeld, Harderup, & Holm, 2007). Since his experiment, indoor moisture content has been correlated with the materials exposed to the indoor air, including the envelope and the furniture.

Moisture plays an important role in creating favourable conditions for mould and biological growth, with potential effects on human health, as a result of the hygrothermal behaviour of buildings, especially in relation to moisture buffering. These effects, however, are yet to be fully investigated. Moisture buffering (MB) is the ability of materials to store and release moisture through a series of absorption and desorption cycles. Hence, building materials are able to attenuate or mitigate variations in indoor RH, helping to maintain its value inside the optimal range (Salonvaara, Ojanen, Holm, Künzel, & Karagiozis, 2004; Simonson, Salonvaara, & Ojanen, 2004). The sorption process also influences the latent heat loads of a building, with clear consequences for HVAC energy consumption. During winter, hygroscopic materials adsorb latent heat, reducing the need for heating (Kraniotis, Nore, Brückner, & Nyrud, 2016), while in summer they reduce the need for cooling by increasing the room's enthalpy (Osanyintola & Simonson, 2006). This characteristic offers a degree of resilience to the indoor environment, which may continue to guarantee thermal comfort for a longer period with minimum contribution from the HVAC system, thereby improving the overall energy efficiency of the building (Nore et al., 2017; Zhang, Qin, Rode, & Chen, 2017). Thus, the impact of moisture buffering extends beyond the material level, exerting considerable influence over energy efficiency and the whole of the indoor environment.

There is as yet no standardised interpretation of the phenomenon, due to both the high level of complexity of the exchange between material and environment, and to the fact that it is a physical quantity that refers simultaneously to material properties and building volume and characteristics (Cascione, Maskell, Shea, & Walker, 2019).

5.1.1 Moisture buffering at different scales

Moisture buffering is a dynamic phenomenon that results from the physical necessity for a system to find a moisture balance between its elements. For example, a completely dry material placed in a humid environment will be subjected to adsorption processes in an attempt to find the equilibrium between the air in the pores and the air in the surrounding environment; hence the moisture will flow from the indoor air to the pores. The opposite process occurs when a wet material is placed in a dry environment, with desorption processes taking place and moisture transferring from the material to the air. Whilst the phenomenon is rooted in the physical laws of moisture exchange in hygroscopic materials, there is currently no agreed framework in the field, and different methods are used to quantify the moisture buffering ability and its impacts at the building level.

A number of different parameters and approaches can be used. For example, in regard to time, moisture buffering can be considered over a short period, such as daily or even seasonal variations. Basically, materials are constantly subjected to hygrothermal exchanges, and these exchanges can be quantified considering different time intervals based on the type and goal of assessment. For example, dynamic hygrothermal assessments aimed at identifying potential condensation or mould risk in the building envelope are performed on an hourly timestep, while other applications may use a different time unit. Similarly, moisture buffering is a phenomenon that involves different spatial scales, from the material properties to the whole building. It is a phenomenon that depends not only on the material characteristics but also on the way in which the material interacts with the adjacent indoor space.

Accordingly, moisture buffering capacity can be assessed in relation to different scales, and different definitions can be used to identify and express this ability to regulate humidity (Fig. 5.1). Although there is currently no officially approved set of definitions to describe these different levels, the ranking system proposed in Svennberg et al. (2007) and Rode, Peuhkuri, Mortensen, et al. (2005) is very widely used.

The first level concerns the material itself and its hygrothermal properties, including moisture capacity, defined as the slope of the sorption isotherm of the specific material (Svennberg et al., 2007). The different properties align to define the capacity of the material to exchange humidity with the air; this is called the ideal moisture buffer value (MBV_{ideal}). It is called *ideal* because it refers to the raw capacity of the material considered only in relation to a thin layer of stagnant air, whose thickness depends on the air velocity within the environment with which the material is exchanging moisture.

When materials are considered in their real application, or assembled to create building components, the moisture buffer capacity is referred to as practical, because it includes surface treatment and layering. It is at this level that the time variable becomes important, as it assesses the capacity of a system in its context, rather than defining the ideal potential of a material.

The last level considers an enclosed space, accounting therefore for its air volume and ventilation, as well as its geometry and mechanical systems. At this scale, the

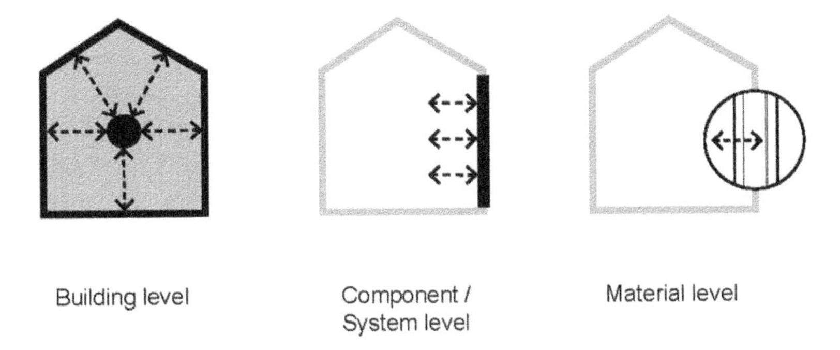

Figure 5.1 Moisture buffering can be classified based on the scale at which the performance are assessed: buildings, system or material.

general term used to refer to moisture uptake and release is moisture buffer performance, although different definitions are used in the literature to quantify this value.

Moving from the first to the third level, there is a loss of agreement on definitions and methods of quantifying the buffering potential, leaving room for different interpretations of the phenomenon, depending on the boundary conditions considered. This lack of standardisation is likely due to the increasing level of complexity and increased specificity of the final performance, with little possibility to generalise, transfer and apply the same results to different contexts.

5.2 Moisture buffering at the material level

Moisture buffering at the material level characterises the material's capacity to absorb and release humidity by interacting with the surrounding environment, which is greatly affected by the material's hygrothermal properties. These basic properties and their intercorrelations are usually nonlinear and moisture- and temperature-dependent, hence they are usually quantified through empirical tests.

The first parameter that describes the capacity of a material to uptake and release moisture is defined by the sorption isotherms, which are expressed for a specific temperature and identify the relation between the water content and the relative humidity of a material. As seen in Chapter 2, Principles of hygrothermal processes, the water content in relation to RH is usually higher during the drying process, creating a hysteresis between sorption and desorption. However, it is also possible to identify a series of scanning curves that interconnect the hysteresis, represented by the dash lines in Fig. 5.2. These scanning lines are created when, during the drying process, the material is re-wet before all pores are completely dry, and water activity follows a path between the wetting and drying boundaries. Hence, there can be an infinite number of scanning curves, depending on when and where the drying process is interrupted. This behaviour determines the process needed for the empirical tests to determine the sorption isotherm without incurring false results due to the detection of a scanning curve. The majority of the methods used to determine the sorption isotherms are based on the equilibration at constant RH, usually achieved through the use of humidity stabilisers, such as salt solutions (Greenspan, 1977), unsaturated salt solutions (Chirife & Resnik, 1984; Clarke & Glew, 1985) and sulphuric acid solutions (Svennberg et al., 2007) or through controlled environment and climate chamber tests (Arlabosse, Rodier, Ferrasse, Chavez, & Lecomte, 2003; Janz & Johannesson, 2001; Ojanen & Salonvaara, 2003). The moisture content in the sample is then measured using gravimetric methods, which rely on the use of a scale to measure the weight of the sample as a proxy for the water content.

From the sorption isotherm it is possible to quantify the moisture capacity of the material, which is defined for a specific interval of relative humidity:

$$\xi = \frac{\Delta w}{\Delta RH} \tag{5.1}$$

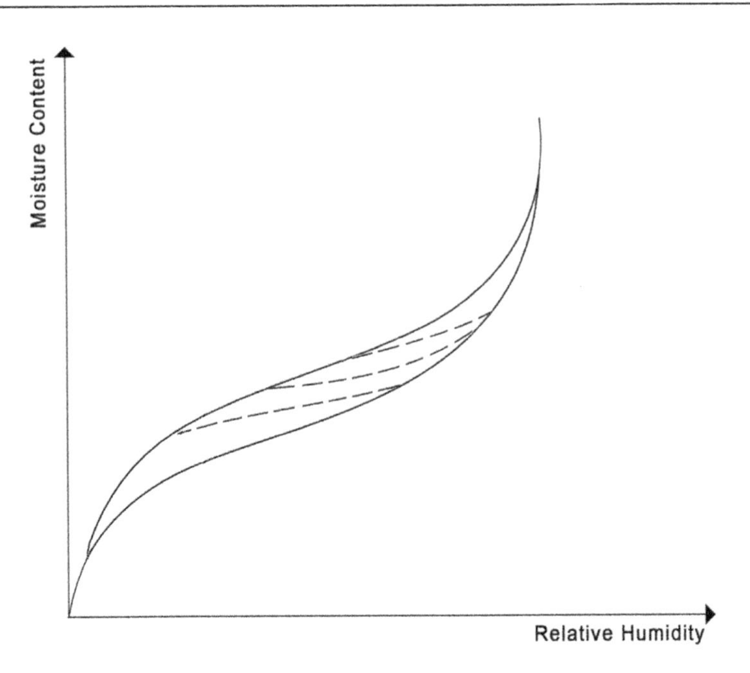

Figure 5.2 The scanning curves indicate possible sorption and desorption alternative paths followed during wetting and drying.

However, this parameter might not include the transitory nature of the MB phenomenon, as the sorption curves are usually drawn by the equilibrium moisture content for each RH and calculated for a specific temperature.

A second important parameter for the moisture buffer evaluation is vapour permeability δ_v. This parameter represents the quantity of water per unit area that can be diffused in a unit of time, calculated for a specific gradient of relative humidity. It is a nonlinear function and usually depends on both ambient RH and time, as ageing of the material surface may impact its hygrothermal properties. Similar to the sorption isotherm, the vapour permeability of different materials is quantified through empirical tests. The traditional method to determine the vapour permeability of a material is the cup-test (Huang & Qian, 2008; McCullough, Kwon, & Shim, 2003), where the specimen is placed as a separator between two environments at different, but known and precisely controlled, relative humidity. The specimen is therefore subjected to a steady-state vapour transfer, which is then determined through gravimetric measurements. The two RH are selected according to the standard referenced for the test. The ASTM E96 standard offers two different scenarios:

- The dry cup method: one value is kept at RH 0% and the other at RH $< 100\%$. In this case it is very common to select 50% as the second RH, as this guarantees it will be in the hygroscopic range of the material, and it allows well-defined values of vapour permeability to be obtained.

- The wet cup method: one value is at RH > 0% and the other is kept at RH 100%. In this case, the side of the specimen in contact with the second space is above its hygroscopic range, and thus the value is a combination of vapour and liquid transfer. However, if there is no formation of condensate, the liquid component is negligible.

To account for nonsteady state variations, different tests have been developed to incorporate additional information about the vapour and liquid transport according to different ambient vapour pressure. This is achieved by varying the ambient RH around the sample or with specific climatic chamber experiments where the vapour released and condensed in the room is measured along with sample weight (Padfield, 1998).

5.2.1 Moisture effusivity and ideal moisture buffer value definition

The sorption curve and vapour permeability are good indicators for characterising the material properties. However, due to the empirical limitations, they cannot fully capture the transitory nature of the buffering phenomenon. Moisture effusivity b_m is a theoretical parameter that can be used to indicate the rate at which moisture is absorbed by a material when subjected to variations in surface humidity (Rode, Peuhkuri, Mortensen, et al., 2005). Moisture effusivity can be calculated as

$$b_m = \sqrt{\frac{\delta_v \cdot p_0 \frac{\partial_v}{\partial RH}}{p_{sat}}} \tag{5.2}$$

where δ_v [kg/(msPa)] is the vapour permeability, p_0 (kg/m^3) the density of the sample, $\partial v/\partial RH$ the relation between the moisture content and the relative humidity, and p_{sat} (Pa) the saturation pressure.

This parameter describes an ideal situation where the surrounding air does not oppose resistance to the moisture exchange, meaning that its convective mass transfer coefficient tends to infinity. Moisture effusivity is a parameter that works for sudden change of surface humidity. However, building materials are usually exposed to cyclical variations in occupancy during the day, with alternating periods of high and low humidity. Considering that the materials tend to stabilise the indoor humidity and find the equilibrium, it is possible to assume that the quantity G_t of moisture absorbed during the high RH period (t_p) is equal to the moisture released during the succeeding low RH period. Applying the Fourier analysis (Rode, Peuhkuri, Hansen, et al., 2005), it is possible to quantify this value as:

$$G_t \approx 0.568 \cdot b_m \cdot \Delta p \cdot \sqrt{t_p} \tag{5.3}$$

where G_t is the moisture uptake (kg), b_m the moisture effusivity, Δp the pressure difference (Pa) and t_p the time period (seconds).

G_t represents the quantity of moisture that a material exchanges with the environment during a given interval of time. The ideal moisture buffering value (MBV_{ideal}) is then defined by normalising the moisture uptake with the variation of surface RH:

$$MBV_{\text{ideal}} = \frac{G_t}{\Delta RH} = 0.00568 \cdot b_m \cdot p_{sat} \sqrt{t_p} \tag{5.4}$$

This value represents the upper limit of the moisture buffer capacity of the material, and is based on moisture effusivity, time period and saturation vapour pressure. MBV_{ideal} aims to define the buffering capacity of a material in a transient condition, but its dependency on the moisture effusivity, defined on the basis of steady-state material properties, may create a discrepancy. Currently, however, there is no empirical method of quantifying MBV_{ideal} so it is difficult to assess possible errors.

5.2.2 Moisture penetration depth

The moisture uptake of building materials relates to hygroscopic range and involves the diffusion of moisture within the pores. When a material is exposed to an environment, diffusion takes place in an attempt to reach equilibrium between the RH inside the pores and in the adjacent environment. Moisture penetration depth represents how deeply moisture can penetrate inside a material for given variations in moisture content in a defined time frame, which is usually set at 24 hours to account for typical variations in buildings. At different time intervals over the 24 hours, the relative humidity distribution inside the material (used as a proxy for the moisture content) will follow a different curve.

RH amplitude, defined as the difference between the maximum and minimum value at each point of the material within the time interval considered, decreases with the depth of the material (Woods, Winkler, & Christensen, 2013). There are different definitions of penetration depth (Cunningham, 1992; El Diasty, Fazio, & Budaiwi, 1993), based on the depth considered as representative. The concept of full penetration depth (Arfvidsson, 1999), which considers the maximum penetration allowed, has considerable potential for use in building applications as it indicates whether moisture can penetrate deeper than the cladding: if "full depth" is higher than the thickness of the cladding, this means that moisture is infiltrating the cladding towards the materials behind. Another definition considers the penetration depth as the point at which the amplitude of RH variation is equal to 1% of the variation found on the material surface (Latif, Lawrence, Shea, & Walker, 2015).

5.3 Moisture buffering at the system level

The ideal moisture buffer value is a good theoretical characterisation of the buffering capacity of a material, but it has several limitations. It is only an approximation of the real value, and discrepancies introduced into the calculation might lead to unreliable results. It also assumes that the thickness of the material is greater than the moisture penetration depth. Similarly, moisture penetration depth is a numerical approximation of the phenomenon which overlooks the impact of the surface resistance to moisture transfer, resulting in an overestimation of the real depth (Maskell, Thomson, Walker, & Lemke, 2018). These simplifications make these definitions impractical for building applications, where moisture buffering involves the entire envelope, which is usually multilayered and composed of materials of different thickness.

To overcome these issues, it is possible to introduce a second moisture buffering value at the system level. This value is a purely empirical quantity, which is obtained through standardised experiments. It accounts for building systems, including surface treatment of the materials, their real thickness and exposure. It describes the amount of moisture absorbed and released per exposed surface area and unit of time when the building component is subjected to RH variations. This moisture buffering value may coincide with the ideal value, but only when one material is considered and its thickness is higher than the moisture penetration depth. In all other cases, this relation between the two values cannot be established.

Since this moisture buffering value requires an empirical approach, different testing protocols have been developed to quantify it. All these methods use the same general process: the sample is preconditioned to reach the hygroscopic equilibrium with the environment, sealed on all but one or two surfaces, exposed to isothermal cyclical step-changes of RH, the moisture transported to and from the sample is gravimetrically quantified, and, finally, the buffering value is evaluated. Fig. 5.3 shows the concept underlying all of these protocols: subjected to cyclical variation of RH, a system absorbs and releases moisture from the environment accordingly. The weight of the sample ranges between two values, and the difference can be considered as the amount of moisture exchanged.

The main differences between the protocols are reflected in the cycles of RH and the final expression of results (Cascione et al., 2019; Kreiger & Srubar, 2019). However, all tests rely on the same set of assumptions and suffer from similar limitations. The testing protocols attempt to replicate, wherever possible, the real application by using boundary conditions, RH cycles and sample sizes that are considered to be representative of the actual building component. A first assumption concerns the sample construction. In typical building construction, a system, or building component, is part of an envelope and, as such, only one face is directly exposed to the interior environment. For this reason, the protocols require sealing of the sides and the exterior face of the sample to prevent them from participating in the moisture exchange and consequently overestimating the overall MB capacity. A second assumption regards the RH step-changes, which follow a 24-hour cycle to represent the daily usage of the building. Although this assumption may represent a standard occupancy profile, it must be noted that in real applications the relative humidity fluctuates constantly, with no stabilisation on a specific value for prolonged time intervals as occurs in the tests, resulting in an overestimation of the final buffering value. Similarly, most tests run in isothermal conditions, allowing

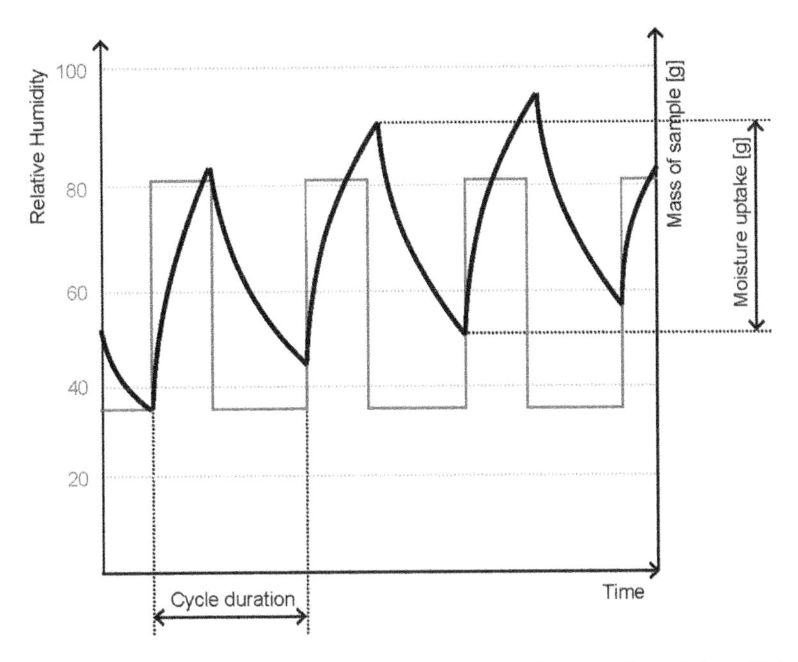

Figure 5.3 The moisture uptake is defined as the difference between the weight at high relative humidity (RH) and low RH (wet and dry).

all the material to achieve a quasi-steady-state equilibrium, and neglect the contribution of the airflow around the sample in the determination of the buffering capacity, which can impact the final value up to 20% (Svennberg et al., 2007). On the other hand, given that all the tests require controlled environmental conditions and are therefore performed in a climate chamber, the sample used is usually a real scale but smaller prototype of the material or building component.

These simplifications are at the basis of the inconsistencies found between the material-level moisture buffering value and the final whole-building moisture buffering performance. Thus the closer the boundary conditions of the testing protocol are to those found in the indoor environment, the more reliable are the results (Roels & Janssen, 2006).

5.3.1 Experimental methods

Over the years, various protocols have been developed in different countries. The most widely used are elaborated below.

5.3.1.1 Step response method DIN 18947

The step response method is the first and simplest laboratory-based testing protocol developed to assess the buffering capacity of materials. This test measures the weight variations of a sample when subjected to an adsorption phase, followed by a desorption phase. The first tests were performed on samples of 15–40 cm, depending on whether the sample material was used for walls or floors, and sealed with paraffin. After the preconditioning phase, which is performed in a climate chamber at 20°C and RH 40% and lasting a few weeks, the samples is exposed to RH 80% and weighed every 3 hours (Kuenzel, 1965). The buffering ability is then expressed through an adsorption coefficient that linked the moisture uptake and the square root of time, calculated from a plot of the moisture uptake. However, this protocol does not provide indications for the desorption phase, failing to standardise the methodology and provide a framework to allow for a direct comparison of the results. A number of similar tests has been developed based on this method (Svennberg et al., 2007), including a German standard developed to specifically address the hygric properties of earthen plasters, where the RH steps were fixed at 50% and 80% in 12 hour cycles (DIN 18947, 2013). In this standard, the moisture buffering ability is expressed as the per cent-mass change.

5.3.1.2 The Padfield method

The step response method adopted a square-wave function to define the RH step changes, which does not represent the real indoor fluctuations observed in buildings. The Padfield method (Padfield, 1998), initially developed in the field of historic building preservation, sought to overcome this problem by reproducing sinusoidal humidity variations: RH ranges between 50% and 60% and the airflow is controlled and kept between 0.2 and 1.2 m/s, to ensure adequate mixing of the indoor air. The method does not directly assess the adsorption property of the material, but calculates the effects of hygroscopic materials on indoor RH by quantifying the difference between the amount of water used during the humidification cycle and the amount recollected during dehumidification. This method is particularly useful for assessing the impact of materials on the indoor environment, but it cannot be used to characterise a building system.

5.3.1.3 JIS A 1470-1 and ISO 24353

Based on the German standard, the Japanese industrial standard A 1470-1 is now one of the most widely used protocols for assessing the moisture buffering capacity of a building system in industrial contexts (JIS A1470-1, 2002). This protocol prescribes a minimum sample size of 100 cm^2 and a material surface film resistance of $4.80 \pm 0.48 \times 10^7$ Pa/kg. The samples are initially preconditioned at a fixed temperature of 23°C and fixed RH that can be selected as 43%, 63% or 83%, and then exposed to 24-hour cycles of RH variations between high and a low humidity levels. The protocol distinguishes between a step signal and a square wave signal, referred to as 'moisture adsorption/desorption test' and 'periodically regular moisture adsorption/desorption test', respectively. In both cases, it is possible to choose the pair of RH values that is deemed relevant, offering three different options: 33% and 53%, 53% and 75% and 75% and 93%. The final buffering value is reported as sorption mass and, for the step signal method, as rate of moisture uptake and release over time.

The ISO 24353 standard is based on the JIS method and was first published in 2008. Similar to the JIS method, the sample size must be 100 cm^2, and samples must be sealed with aluminium tape on all surfaces except one. The protocol prescribes a fixed temperature for the preconditioning phase, 23°C, and allows choice of the RH as 43%, 63% or 83%. The RH testing follows 12-hour cycles, and allows the selection of the RH pair as 33% – 53%, 53% – 75% and 75% – 93%. (ISO 24353, 2008).

5.3.1.4 The NORDTEST

The NORDTEST protocol (Rode, Holm, & Padfield, 2004; Rode, Peuhkuri, Hansen, et al., 2005) is the most commonly used method for assessing the moisture buffering ability of building system in research, since it was developed with the specific purpose of creating an accepted standardised approach to assessment. One of the main reasons for its widespread use is related to the validation process it underwent, which was conducted through a round robin MBV procedure in different testing centres, and which showed good agreement on the final results (Rode, Peuhkuri, Mortensen, et al., 2005). This protocol requires samples to be squared and have a surface of 100 cm^2, while the thickness must be greater than the theoretical moisture penetration depth of the material. Similar to the JIS method, the experiments are performed at a constant temperature of 23°C. Samples undergo a preconditioning phase at 50% RH until they reach hygroscopic equilibrium and are then successively exposed to RH variation cycles of 16 hours at 33% RH and 8 hours at 75%. These steps and time intervals are defined based on the typical occupancy pattern of commercial spaces, where higher levels of moisture generation are observed during working hours, followed by a longer period at lower humidity. The NORDTEST specifies the method for keeping the humidity constant, allowing both the use of a climate chamber and of salt solutions for small-scale tests.

The final moisture buffering ability is expressed as Moisture Buffer Value practical. It is reported by normalising the uptake value G (g) per exposed surface area A (m^2) and change in relative humidity ΔRH (%):

$$MBV_{\text{practical}} = \frac{G}{A \cdot \Delta RH} \tag{5.5}$$

Based on this definition, a later index was developed to account for different cycles of high and low humidity. In the case of a residential building, the 8–16 hours alternation may not be representative of the real occupation pattern. This index is an implementation of the $MBV_{\text{practical}}$ and is expressed as (Janssen & Roels, 2007):

$$MBV^* = \alpha MBV_{8h} + (1 - \alpha) MBV_{1h} \tag{5.6}$$

The weighting factor α depends on the production interval and can be interpolated from the values below:

- Short production interval (0–2 hours): 0
- Medium production interval (2–6 hours): 0.5
- Long production interval (6–10 hours): 1

5.3.1.5 Ultimate moisture buffering value

Based on the NORDTEST protocol, this method accounts for different temperature and humidity values that could be more representative of the seasonal variations usually found in buildings. After the usual preconditioning phase at 23°C and 50% RH, the samples are exposed to extreme conditions representing summer (40°C and 93% RH) and winter (18°C and 3%). The Ultimate Moisture Buffer Value is then expressed as:

$$UMBV = \sum_{i=I}^{III} \alpha_i \cdot MBV_i \tag{5.7}$$

where α is the time coefficient, or total hour/day each sample is subjected to the specified RH (Wu, Gong, Yu, & Huang, 2015). The time coefficient depends on the stage of the test and is equal to 12 hours for the preconditioning, 8 hours for the high humidity step and 4 hours for the low humidity step. The UMBV represents a cumulative value over the three steps and is more indicative of yearly capacity, as it accounts for seasonal variations.

5.3.2 Limitations of the protocols

Table 5.1 shows the main testing protocols used to assess the moisture buffering ability of building systems, which are further explained below.

Among the various protocols, the NORDTEST is the most widely used, so it is more common to see materials characterised using the Moisture Buffering Value practical. After an initial period in which the NORDTEST protocol was tested to quantify the $MBV_{practical}$ of different materials, researchers focused on the parameters that might affect this value in both the test procedure and real-environment situations (McGregor, Heath, Fodde, & Shea, 2014). In fact, multiple factors influence the moisture buffer capacity of a building component. The major issue is the isothermal condition of the test protocols, which do not acknowledge the temperature variations to which buildings are subjected (Zhang et al., 2017). Larger samples or real build-up prototypes can be used to partially resolve this issue (Belarbi, Qin, Aït-Mokhtar, & Nilsson, 2008; Pavlík & Černý, 2009; Yang, Vera, Rao, Ge, & Fazio, 2007), but this solution is difficult to implement as it requires specific equipment that is rarely available for standard building design. Another issue identified in the experimental protocol relates to the moisture load profile and the RH cycles, which can influence the overall final value (Ge, Yang, Fazio, & Rao, 2014). When

Table 5.1 Description of the main testing protocols to assess the moisture buffering ability of building systems.

Protocol	Temperature	Preconditioning	Cycles	Humidity
Step-response	20	40%		80%
JIS A 1470-1	23	43%, 63% or 83%	12 hours	33%–53%, 53%–75%, 75%–93%
NORDTEST	23	50%	16 hours at low humidity and 8 hours at high humidity	53%–75%
ISO 24353	23	43%, 63% or 83%	12 hours	33%–53%, 53%–75%, 75%–93%
UMBV	23 at preconditioning, variable during the test	50%	12 hours preconditioning, 8 hours high humidity, 4 hours low humidity	98% (40°C)–3% (18°C)

shorter RH cycles are applied, the impact of the surface treatment and coating of the sample is greater than in longer cycles (Zhang, Yoshino, & Hasegawa, 2012). Further, all protocols assume air movement, air speed, temperature and the surface resistance coefficient constant during the experimental phase. A few researchers have sought to assess the influence of these assumptions and have found that convective moisture flow greatly impacts the MBV (Allinson & Hall, 2012). However, no currently available protocols have been able to overcome these limitations, suggesting that the real moisture buffering value is highly application-specific and that the value that can be evaluated through the tests must be taken as an indication rather that an actual quantification.

5.3.3 Ranking of the moisture buffer capacity of building systems

The number of different methods available and the different definitions used to assess the moisture buffering ability of building components introduce a level of complexity to comparisons between different materials and/or systems. To date, the NORDTEST protocol is the most often used in the literature as the reference method.

Building component design accounts for climate, specific regulations and building codes, as well as socio-cultural contexts. Hence, it is highly difficult, if not impossible, to identify a set of universal construction practices or build-ups. For one typology of building component, such as brick veneer, there are infinite variations. For example, it is possible to consider only the painting layer and differentiate between bricks that are painted and bricks that are not painted, between latex-based or oil-based paint and different thicknesses of paint. This variation extends to all layers of the component, resulting in a significantly high number of potential combinations. For this reason, the results of a comparison at the building component level are highly case-specific, with limited potential for cross-transference of the knowledge gained. On the other hand, a comparison at the material level may allow material typologies to be grouped into moisture buffering classes, which would give a general indication of comparative performance. For this reason, the system classification is usually performed at the material level, accounting for real world application and considering the potential of the material as an indoor humidity regulator when applied as interior cladding. While it is interesting to have a basis for comparison, there is ongoing discussion about the added value of the MBV tests: when the results are so dependent on the protocols, questions can be raised about their applicability and transferability in the design process as well.

It is possible to identify a general trend in the published research the moisture buffering value. The first studies focused on traditional building materials, such as concrete, cement-based materials and plasterboards (Cerolini, D'Orazio, Di Perna, & Stazi, 2009; Hameury, 2005; Osanyintola, Talukdar, & Simonson, 2006), as well as insulation materials such as mineral wool or cellulose (Pavlík, & Černý, 2008, 2009; Stazi, Tittarelli, Politi, Di Perna, & Munafò, 2014; Toman, Vimmrová, & Černý, 2009). In the second period, particular attention was paid to the definition of the $MBV_{\text{practical}}$ of multilayered building components (Latif, Ciupala, Tucker, Wijeyesekera, & Newport, 2015; Latif, Lawrence, et al., 2015; Latif, Lawrence, Shea, & Walker, 2018) and innovative materials, with the aim of demonstrating their superiority to traditional materials. These include hemp concrete (Colinart, Lelievre, & Glouannec, 2016; Collet & Pretot, 2012; Collet, Chamoin, Pretot, & Lanos, 2013; Ouméziane et al., 2011; Rahim et al., 2015; Tran Le, Maalouf, Mai, Wurtz, & Collet, 2010), bio-based insulation (Nguyen, Grillet, Bui, Diep, & Woloszyn, 2018; Palumbo, Lacasta, Holcroft, Shea, & Walker, 2016) and unfired clay materials (Liuzzi, Hall, Stefanizzi, & Casey, 2013).

As mentioned above, the potential of the moisture buffering value lies in the possibility of comparing different materials and systems, rather than looking at the absolute number alone. For this reason, a ranking system has been introduced to allow the materials (and systems) to be classified on the basis of their ability to uptake and release moisture. This classification is particularly interesting for building applications as a way of indicating the potential of specific hygroscopic materials to impact indoor humidity and, therefore, perceived environmental quality (Cerolini et al., 2009; Fang, Clausen, & Fanger, 1998).

Table 5.2 List of typical building materials and their associated classification and moisture buffer value (MBV).

Material	MBV (g/RH/m^2)	Classification
Bamboo fibreboard	3.0	5
Brick	0.48	2
Compressed earth bricks	1.9	4
Compressed earth bricks with barley straw	2.6	5
Cellulose	3.1	5
Cement	0.37	2
Ceramic	0.26–0.95	1–3
Clay plaster	1.7	4
Concrete	0.7–0.88	3
Earth plaster	2.9	5
Gypsum	0.33–1.1	2–4
Hemp concrete	1.89	4
Hemp fibre	3.3	5
Hemp lime	2.3	5
Laminated wood	0.46	2
Plywood	1.1	4
Spruce board	1.2	4
Spruce plywood	0.57	3
Fibreboard	2.4	5

The classification divides the MBV into five categories:

1. Negligible effect: MBV below 0.2 g/m^2/RH
2. Limited effect: MBV between 0.2 and 0.5 g/m^2/RH
3. Moderate effect: MBV between 0.5 and 1 g/m^2/RH
4. Good effect: MBV between 1 and 2 g/m^2/RH
5. Excellent effect: MBV above 2 g/m^2/RH

In relation to the materials listed above, it can be observed that, generally, traditional materials have negligible $MBV_{\text{practical}}$ compared to bio-based materials. Concrete can be categorised as 2, gypsum as 3 and bricks between the two. Hemp concrete ranges between category 4 and 5, cellulose and bio-based insulation are in category 5 and unfired clay bricks exceed the expectation of category 5. These results clearly show that bio-based materials are highly hygroscopic and have better moisture buffering properties. Table 5.2 shows the practical moisture buffering values of common building materials and their classification.

5.4 Moisture buffer performance at the building level

When moving from the system to the room scale, the level of complexity embedded in the assessment and definition of moisture buffering capacity increases further. All the testing protocols for assessment at the system level assume constant air speed, temperature and surface resistance coefficient, which is far from the actual conditions found in real buildings. Further, a number of other parameters greatly influence the moisture buffer performance of a room including, but not limited to, the volume of the space, air infiltration, the presence of doors and windows and the furniture (Kurnitski, Kalamees, Palonen, Eskola, & Seppänen, 2007). The moisture buffering performance of a room must therefore be assessed considering its whole context and replicating the real conditions of use to assure a reliable result.

Different approaches have been developed to test, quantify and validate moisture buffering performance. They can be divided into in situ measurements, climate chamber experiments and virtual simulation assessments. Each of these has different strengths and weaknesses, and their use must be carefully selected based on the scope of the analysis. For example, hygrothermal simulations are a quick and relatively nonexpensive method suitable for a comparative analysis at the design stage. However, they suffer from the same limitations that apply to all virtual simulations, such as errors produced by the software and the broad number of simplifications required. In situ measurements are probably the most precise and accurate quantification method, as they give a snapshot of the real behaviour of the building under the real conditions of use. However, they can be destructive or invasive for the occupants, and are more suitable for retrofit interventions after the building has operated for a certain period of time. Full-scale climate chamber experiments are considered necessary to validate hygrothermal simulations and laboratory scale tests, but they are very expensive, time consuming, and require specific facilities that can accommodate a real-scale mock-up inside a climate room.

All these approaches lack an agreed methodology that could unify and standardise the results to enable a direct comparison of the different cases. This issue highlights both the complexity of moisture buffering performance evaluation and the importance of finding a method that can link material properties to systems and to whole building performance in order to better understand the phenomenon and its implications in real applications.

5.4.1 *Definitions of moisture buffering performance*

Just as a number of different protocols are available to quantify the MBV at the systems level, different methodologies and parameters are used to quantify moisture buffering at the building level. The wide range of approaches available to quantify moisture buffering performance is reflected in the variety of definitions and calculation methods for this value. Unlike the moisture buffering value practical, no definition is more common than another, and it is even more common to find studies where no definition at all is provided. The majority of studies focus on the assessment and evaluation of the effects of moisture buffering on the indoor environment and energy consumption, rather than trying to create an indicator for it. For this reason, the following definitions are to be considered as an indication of the variability of approaches. This issue highlights the need for a coherent body of research to establish a clear assessment method.

5.4.1.1 Accumulated moisture value in the surface

The definition of the accumulated moisture value (Li, Fazio, & Rao, 2012) is based on the measurement of all the parameters that determine the hygrothermal exchange of a room. It requires a full-scale climate chamber approach, where the air handling unit can weigh the condensed water. The value accounts for air infiltration, moisture diffusion through the envelope, and moisture removed through ventilation.

$$M_{mat}(t) = \left[-M_a(t) - M_{diff}(t) - M_{vent}(t) + G(t)\right] \tag{5.8}$$

where M_a is the moisture change in the room, M_{diff} is the vapour diffusion through the walls, M_{vent} is the vapour removed by ventilation and G is the moisture source.

5.4.1.2 Effective damp relative humidity

Effective damp relative humidity (EDRH) (Li et al., 2012) is a parameter based on the accumulated moisture value. It was developed to simplify and further refine the calculation process. This index accounts for nonhomogenous moisture distribution inside the climate chamber and requires two test conditions: one with real building components, the other with fully nonhygroscopic building components, achieved through the use of moisture impermeable material – 0.8 mm thick aluminium metal sheets – to cover the interior lining. The EDRH expresses the maximum difference between the average RH value found in the test room (monitored by sensors) under the two different scenarios.

$$EDRH = \frac{p_t \cdot \Delta w_{\max}}{p_{sat}(0.622 + \Delta w_{\max})} \tag{5.9}$$

with p_t barometric pressure, p_{sat} vapour saturation pressure at 21°C and $\Delta w_{\max}$ calculated as:

$$\Delta w_{\max} = \frac{1}{V \cdot \rho_{\text{air}}}\left[\left(\sum_{j=1}^{k} \rho_{\text{air}} \cdot v_j \cdot \chi_j^n(t) - \sum_{j=1}^{k} \rho_{\text{air}} \cdot v_j \cdot \chi_{ij}^n(o)\right) - \left(\sum_{j=1}^{k} \rho_{\text{air}} \cdot v_j \cdot \chi_j(t) - \sum_{j=1}^{k} \rho_{\text{air}} \cdot v_j \cdot \chi_{ij}(o)\right)\right] \tag{5.10}$$

where ρ_{air} is the air density, V the volume of the room, j is the jth sensor, χ the humidity ratio, v_j the volume around the sensor (equal to 1/33 of the room volume) and k the number of sensors.

5.4.1.3 Hygroscopic inertia

The hygroscopic inertia of a room has been developed from climate chamber experiments and aims to represent a similar concept to that of thermal inertia. It seeks to correlate the daily RH variation to the hygroscopic level of the room, meant as the superposition of the moisture buffering value of all the hygroscopic interior finishes and objects (Ramos & De Freitas, 2006; Ramos & De Freitas, 2009):

$$HIR = \frac{\sum A_k \cdot MBP_k + \sum MBP'_l}{V} \tag{5.11}$$

where HIR (kg/m^3/%RH) is the hygric inertia per cubic metre of room, MBP_k (kg/m^2/%RH) and A_k (m^2) the moisture buffer potential and area of finish k, MBP'_l (kg/%RH) the equivalent moisture buffer potential of object l and V (m^3) the volume of the room.

This concept has been further improved and generalised to the production-interval moisture buffering value (Vereecken, Roels, & Janssen, 2011):

$$HIR = \frac{\sum A_k \cdot MBP_k^* + \sum MBP'_l*}{V} \tag{5.12}$$

where HIR 8 h/1 h (kg/m^3/%RH) is the long/short term inertia.

Numerical simulations performed for different material sand thicknesses showed good agreement between the RH balancing and the HIR* value (Janssen & Roels, 2009). This indicates that there is no need for a full-scale test to assess the hygroscopic inertia of the room since but it can be evaluated starting from the materials MBV.

5.4.2 Influence of moisture buffering performance on hygrothermal behaviour of buildings

The overall capacity of a building to regulate moisture is determined not only by its walls but also by all the hygroscopic materials that are in contact with the indoor air, including furniture and objects such as books and textiles. From this perspective, the moisture buffering performance of a room may be much higher than what can be calculated by assessing only the contribution of the walls. It has been estimated that the quantity of moisture involved in the buffering from walls and furniture can be equal to the quantity released by a person occupying the space (Tran Le et al., 2016), indicating the potential of hygroscopic materials to passively regulate indoor humidity. This ability is key to ensuring a healthy indoor environment.

Indoor humidity is a parameter that is often overlooked, even though its implications for human health are widely recognised. The indoor RH should be kept between 40% and 70%, which is considered the optimal range for health. Some studies show that dry air can lead to skin and respiratory irritation (Reinikainen & Jaakkola, 2003), while high RH levels are linked to asthma, wheezing and bronchial hyperresponsiveness (Bornehag et al., 2001). The smart use of high moisture buffering materials can contribute to maintain humidity in this range and assure healthy indoor environments by reducing the risks of mould and condensation (Bairi, Gomez-Arriaran, Sellens, Odriozola-Maritorena, & Perez-Iribarren, 2017). Materials with high moisture buffering value can mitigate humidity variations by reducing the maximum indoor RH by up to 20% (Simonson et al., 2004).

The influence of moisture buffering on humidity not only increases the overall quality of the indoor environment but also, as a result, it impacts overall energy consumption of the building. During summer, hygroscopic materials lower the indoor humidity, decreasing the room enthalpy (Osanyintola & Simonson, 2006), which can reduce the average cooling needs by between 4% (Moon, Ryu, & Kim, 2014) and 30% (Zhang et al., 2017). During winter, in contrast, the adsorption of moisture from the air generates latent heat (Kraniotis et al., 2016) that results in a reduction in heating consumption of around 15% (Woloszyn, Kalamees, Abadie, Steeman, & Sasic Kalagasidis, 2009). These effects are more significant when low air change rates are specified for the mechanical ventilation, as the ventilation itself is not able to reduce the moisture load and the contribution of the materials' adsorption becomes more significant (Woloszyn et al., 2009; Zhang et al., 2017). However, moisture buffering is rarely considered in the design of a building's heating, cooling and ventilation system, despite the evidence on the positive effects of MBV.

In short, it is important to consider the passive ability of materials to absorb and desorb moisture in estimating the peak heating and cooling demands in both residential and commercial buildings. It is therefore clear that the moisture buffer performance of a building can no longer be neglected during the design stage, but must be considered and assessed to guarantee reliable and efficient humidity management.

5.5 Prediction of the moisture buffering effects on a building's hygrothermal behaviour

Since moisture buffering has significant effects on the hygrothermal behaviour of buildings, it is important to be able to quantify its contribution to indoor humidity during the design phase. An accurate and reliable prediction of the indoor humidity is key to ensuring adequate indoor hygrothermal comfort and correct design of the HVAC system. Nonetheless, in whole building simulations software, which was initially developed for thermal-only calculation, moisture buffering is often treated as an independent variable. It is therefore not included in the moisture transport equations but, rather, is evaluated separately through a different mathematical model. This introduces a source of error that prevents the full potential of moisture buffering effect being accounted for in numerical simulations (Cascione et al., 2019).

The moisture uptake and release of hygroscopic materials influence the final moisture balance of a room, which can be expressed as:

$$\frac{V}{R_v \cdot T_i}\frac{\partial p_i}{\partial_t} = (p_e - p_i)\frac{n \cdot V}{3600 \cdot R_v \cdot T_i} + G_{vp} - G_{buff} \tag{5.13}$$

with V volume of the room (m^3), R_v gas constant, T_i indoor temperature, V/R_tT_i indicates the moisture capacity of the room, $p_{e/i}$ the partial vapour pressure for indoor and outdoor air (Pa), G_{vp} and G_{buff} the moisture produced and buffered (kg/s).

This equation is typically used in building simulation software. It assumes an ideal convective mixing of the air (that is, temperature, humidity and pressure are equal in all points of the space), no surface condensation and no dependency of air density on temperature (Janssen & Roels, 2009). The moisture buffered can be rewritten as function of the vapour diffusion:

$$G_{buff} = \sum\nolimits_k \beta_k \cdot A_k \cdot \left(p_i - p_{s,k}\right) \tag{5.14}$$

with β_k (s/m) the convective surface film coefficient for vapour transfer, A_k (m^2) the surface area, $p_{s,k}$ (Pa) the surface partial vapour pressure for the element k.

To assess the variations in indoor humidity, the room balance must be solved simultaneously with the equation that represents the moisture equilibrium inside the materials. Hence, a third equation must be introduced to represent the mass transport in each k of those:

$$\frac{\partial w_k}{\partial t} = \frac{\xi_k}{p_{sat}} \cdot \frac{\partial p_{v,k}}{\partial t} = \nabla\left(\delta_k\left(p_{v,k}\right) \cdot \nabla p_{v,k}\right) \tag{5.15}$$

with w_k moisture content (kg/m^3), ξ_k the moisture capacity (kg/m^3), p_{sat} the saturation vapour pressure (Pa) and δ_k the water vapour permeability of the absorbing material k.

Note that the latter is nonlinear, as vapour permeability and moisture capacity are humidity-dependent values. The set of equations can be solved numerically, but the complexity embedded in the calculation makes it difficult for it to be applied to multilayered or multidimensional building elements, which would require a significant computational and experimental load to be solved (Janssen & Roels, 2009). For this reason, different simplified models have been developed to account for the materials' moisture buffering contribution.

The first model developed to account for moisture buffering of building materials is the so-called effective capacitance method (EC) (Yang, Fu, & Qin, 2015), which links the capacity of the internal finishing and the room air to store moisture by introducing an air capacitance multiplier. The EC model does not really solve the moisture transfer issue by implementing the materials' moisture buffering ability but, rather, focuses on describing its effects on the indoor air. The underlying idea is to increment the air volume and, hence, its overall moisture buffering capacity, to account for the additional moisture uptake and release from the hygroscopic materials. This model assumes that the humidity stored in the material is in equilibrium with the air humidity at any time. Based on this, the moisture uptake and release equation can be rewritten as:

$$M \cdot \frac{V}{R_v \cdot T_i} \cdot \frac{\partial p_i}{\partial t} = (p_e - p_i)\frac{n \cdot V}{3600 \cdot R_v \cdot T_i} + G_{vp} \tag{5.16}$$

where M is the multiplier factor, which can range from 10 to 25, depending on the type of building (e.g. offices, library, apartment) (Fang, Winkler, & Christensen, 2011; Henderson, Parker, & Huang, 2000). The higher the value of M, the more the room is able to buffer humidity. Although this model significantly simplifies the computation of indoor humidity, it is a purely empirical method that does not represent the reality; indeed, the lumping process implies that the surface resistance and the vapour permeability are negligible. The exclusion of the time-dependent behaviour of moisture storage leads to the inability of the model to predict the humidity profile over time. However, it allows an estimation within reasonable ranges of the maximum and minimum values, which are the parameters used for a correct HVAC design.

If prediction of the humidity over time is required, a second model can be used to define it more precisely, namely, the effective moisture penetration depth (EMPD) model (Cunningham, 2003). This method has been developed based on the discretisation of the differential equations of the moisture transfer within the materials. It assumes that the moisture transfer takes place between the indoor air and a thin layer of material characterised by uniform moisture content (Kerestecioglu, Swami, & Kamel, 1990). Assuming that temperature and vapour pressure vary linearly in this layer, its thickness depends on the cyclical variations of humidity that characterise the building application:

$$d_b = a \cdot d_p = a\sqrt{\frac{t_p \cdot \delta \cdot p_{sat}(T_b)}{\pi \cdot \xi}} \cdot \alpha = \min\left(\frac{d}{d_p}, 1\right) \tag{5.17}$$

where t_p is the period (s) and a is a correction factor that accounts for the fact that the actual thickness d (m) may be lower than the moisture penetration depth d_p.

The EMPD method can be developed in two different formulations, depending on whether the nonisothermal or isothermal model is used (Woods et al., 2013). This method accounts for shorter cycles of RH variations and assumes that moisture storage and buffer involve only one layer. A second formulation allows the inclusion of deeper moisture buffering processes due to longer RH variation cycles (Woods et al., 2013). It models the moisture transfer from the air to the surface layer and from the surface layer to the deeper layer. In either case, the EMPD is more accurate than the EC model, and it can find the exact solution when the humidity variation is perfectly cyclical, or a good and realistic approximation when it is not. However, this model, in both variations, relies heavily on the moisture penetration depth value, which requires time and labour-intensive measurements, given its dependency on the moisture capacity and vapour permeability. This method fails to provide a framework to account for objects and furniture, which greatly contribute to the final buffering performance, and does not allow the superposition in one general summative buffering capacity of the contribution of the different materials, which would significantly simplify its application (Janssen & Roels, 2009).

The refined heat, air and moisture (HAM) model for moisture transfer has also been used to develop finite-difference models to predict the storage of moisture in hygroscopic materials; an example is the heat, air and moisture transfer (HAMT) model. These models are more accurate and sophisticated than the EC and EMPD, and are at the basis of the numerous forms of hygrothermal software described in Chapter 6, Hygrothermal modelling. A drawback of these approaches is the significant computation time required for application to whole building simulations, as well as the numerous material properties that are needed as input. Further, most of these software programs are designed to solve moisture transfer as a one-dimensional issue, therefore they look at the system level rather than the performance level.

5.6 Moisture buffering-aware design

Moisture buffering and hygroscopic materials have a significant impact on a building's hygrothermal performance and, despite substantial research efforts over years, this remains an underdeveloped field that offers numerous opportunities for further development and improvement. From the design point of view, a clear and agreed framework on how to account for moisture buffering from the early design stage and, hence, assess its implications, is still lacking. However, some innovative examples can illustrate the huge potential of a MB-aware design, from the material to the whole building scale.

Novel engineered materials, such as superabsorbent hydrogel, have a significantly higher moisture buffering value in relation to their exposed surface compared to traditional or even natural ones (Kreiger & Srubar, 2019). If integrated into the walls or partitions, their improved ability to exchange moisture with the indoors may have significant benefits to the hygrothermal performance of the building (Gianangeli, Di Giuseppe, & D'Orazio, 2017). Another trend in the field of MB-enhanced materials is the use of biotic materials, such as lichens and moss. Cyanobacteria-based gel lichens can absorb 2000% moisture compared to their dry weight (Esseen, Rönnqvist, Gauslaa, & Coxson, 2017; Gauslaa, 2014). In addition to high MBV, their benefits include carbon storage and sequestration, volatile compound sequestration, and the ability to indicate toxic levels of indoor pollutants (Easton, 1994), making them particularly interesting for building applications. Despite the promising results obtained in experimental programs, there is ongoing discussion about the feasibility of such solutions at a larger scale. Indeed, the simplest, cheapest, most effective and already available first step toward a real MB-aware design is informed materials selection during the design process.

The MBV of a material is defined as the quantity of moisture transported to and from a material when it is subjected to cyclical variations of relative

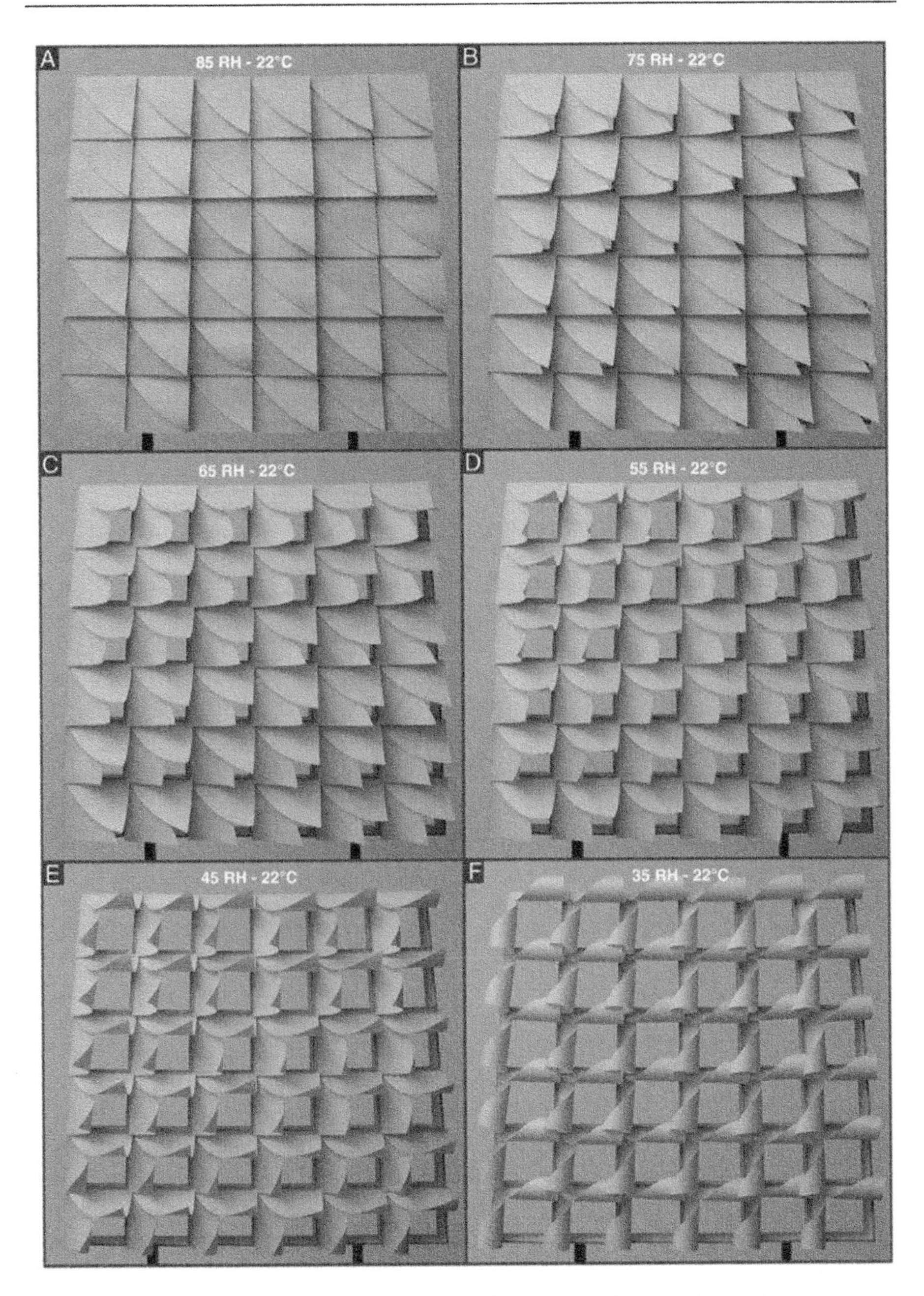

Figure 5.4 Image showing the laboratory tests of element under controlled climatic conditions. Changes in RH coupled with appropriate material quality controls during fabrication enable a homogeneous opening and closing motion across the array. From fully closed (A) to fully open (F), the veneer performs various formal states as moisture is removed from the air.

humidity in relation to its exposed surface. When walls are considered as systems, the principal parameters involved are, therefore, the relative humidity uptake and the buffering surface exposed to indoor air. A second focus of MB-aware design concerns improvement of the overall moisture buffering performance by increasing the exposed surface of hygroscopic materials within a room, taking advantage of the material properties. For example, the capacity of timber to expand and shrink under hygrothermal stimuli has been used to develop objects or panelling systems that respond to humidity variations by changing their configuration and increasing their exposed surface when needed (Vallati, Brambilla, Spalletti, Frigerio, & Ranzi, 2018).

The responsiveness of timber to hygrothermal stresses has also been investigated to generate MB-aware design of whole building components and architectural pavilions. Using a special bi-layer timber thin element, it is possible to create an architectural element that changes its shape with variations in humidity, as shown in Fig. 5.4 (Reichert, Menges, & Correa, 2015). These prototypes rely on the hygromorphic response of timber which, under humidity stress, tends to change its shape. The final response depends on the different dimensional changes of the two layers: the greatest responsiveness is achieved when a highly hygroscopic timber essence, used in the direction tangential to the fibres, is coupled with a nonhygroscopic material (i.e. thin metal foil). This concept has also been applied as a sustainable design feature for climate-responsive facades to optimise daylight by opening in humid weather conditions and overcast skies to allow more light to penetrate, and closing in dry and sunny conditions to provide more shading. (Holstov, Morris, Farmer, & Bridgens, 2015). These research studies demonstrate that moisture buffering of hygromorphic materials provides an opportunity to design buildings that passively adapt to the indoor and outdoor hygrothermal variations. The latest trends in climate-responsive architecture relies on the use of sophisticated technologies, sensors, control systems and actuators. As a result, such buildings are highly energy intensive and have expensive envelopes that often require a high level of maintenance and are difficult to repair (Loonen, Trčka, Cóstola, & Hensen, 2013).

5.7 Concluding remark

The moisture buffering of building material has been a topic of extensive discussion since the early 1970s. However, there is still no common agreed framework in terms of model, methodology and terminology. The moisture buffer value framework is the most widely used, but the terminology used means that confusion about the different levels is often present. The biggest problem is the range of difficulties embedded in the prediction of the moisture buffering contribution to energy consumption and indoor comfort. Efforts are ongoing to identify a reliable model that can be employed during the design stage to quantify the implications of hygroscopic materials for the design and performance of a building.

Based on current knowledge in the field, it is reasonable to predict that future work will aim to consolidate and validate the models through empirical experiments and to extend the boundary of the modelling system to achieve a more reliable and comprehensive tool. Further, additional material characterisations of innovative, sustainable and nonconventional materials are needed to create a shared database that can be used by building designers to inform the design process.

To conclude, new opportunities and future research directions are emerging to advance technical understanding, material development and computational modelling of moisture buffering effects and to leverage the benefits in the design, construction and operation of residential and commercial buildings.

References

Allinson, D., & Hall, M. (2012). Humidity buffering using stabilized rammed earth materials. *Proceedings of Institution of Civil Engineers: Construction Materials*, *165*(6), 335–344. Available from https://doi.org/10.1680/coma.11.00023.

Arfvidsson, J. (1999). A new algorithm to calculate the isothermal moisture penetration for periodically varying relative humidity at the boundary. *Nordic Journal of Building Physics*, 2.

Arlabosse, P., Rodier, E., Ferrasse, J. H., Chavez, S., & Lecomte, D. (2003). Comparison between static and dynamic methods for sorption isotherm measurements. *Drying Technology*, *21*(3), 479–497. Available from https://doi.org/10.1081/DRT-120018458.

Bairi, A., Gomez-Arriaran, I., Sellens, I., Odriozola-Maritorena, M., & Perez-Iribarren, E. (2017). Hygroscopic inertia influence on indoor environments: moisture buffering.

Belarbi, R., Qin, M., Aït-Mokhtar, A., & Nilsson, L. O. (2008). Experimental and theoretical investigation of non-isothermal transfer in hygroscopic building materials. *Building and Environment*, *43*(12), 2154–2162. Available from https://doi.org/10.1016/j.buildenv.2007.12.014.

Bornehag, C., Bonini, S., Custovic, A., Kunzli, N., Malmberg, P., Matricardi, P., & Sundell. (2001). Dampness in buildings as a risk factor for health effects, EUROEXPO-Multidisciplinary review of the literature (1988–2000) on dampness and mite exposure in buildings and health effects.

Cascione, V., Maskell, D., Shea, A., & Walker, P. (2019). A review of moisture buffering capacity: from laboratory testing to full-scale measurement. Construction and building materials. *Construction and Building Materials*, *200*, 333–343.

Cerolini, S., D'Orazio, M., Di Perna, C., & Stazi, A. (2009). Moisture buffering capacity of highly absorbing materials. *Energy and Buildings*, *41*(2), 164–168. Available from https://doi.org/10.1016/j.enbuild.2008.08.006.

Chirife, J., & Resnik, S. L. (1984). Unsaturated solutions of sodium chloride as reference sources of water activity at various temperatures. *Journal of Food Science*, *49*(6), 1486–1488. Available from https://doi.org/10.1111/j.1365-2621.1984.tb12827.x.

Clarke, E. C. W., & Glew, D. N. (1985). Evaluation of the thermodynamic functions for aqueous sodium chloride from equilibrium and calorimetric measurements below 154°C. *Journal of Physical and Chemical Reference Data*, *14*(2), 489–610. Available from https://doi.org/10.1063/1.555730.

Colinart, T., Lelievre, D., & Glouannec, P. (2016). Experimental and numerical analysis of the transient hygrothermal behavior of multilayered hemp concrete wall. *Energy and Buildings*, 1–11. Available from https://doi.org/10.1016/j.enbuild.2015.11.027.

Collet, F., Chamoin, J., Pretot, S., & Lanos, C. (2013). Comparison of the hygric behaviour of three hemp concretes. *Energy and Buildings*, *62*, 294–303. Available from https://doi.org/10.1016/j.enbuild.2013.03.010.

Collet, F., & Pretot, S. (2012). Experimental investigation of moisture buffering capacity of sprayed hemp concrete. *Construction and Building Materials*, *36*, 58–65. Available from https://doi.org/10.1016/j.conbuildmat.2012.04.139.

Cunningham, M. J. (1992). Effective penetration depth and effective resistance in moisture transfer. *Building and Environment*, *27*(3), 379–386. Available from https://doi.org/10.1016/0360-1323(92)90037-P.

Cunningham, M. J. (2003). The building volume with hygroscopic materials—an analytical study of a classical building physics problem. *Building and Environment*, *38*(2), 329–337. Available from https://doi.org/10.1016/S0360-1323(02)00061-6.

DIN 18947. (2013). *Earth plasters—Terms and definitions, requirements, test methods*.

Easton, R. M. (1994). Lichens and rocks: a review. *Geoscience Canada*, *21*(2), 59–76.

El Diasty, R., Fazio, P., & Budaiwi, I. (1993). Dynamic modelling of moisture absorption and desorption in buildings. *Building and Environment*, *28*(1), 21–32. Available from https://doi.org/10.1016/0360-1323(93)90003-L.

Esseen, P. A., Rönnqvist, M., Gauslaa, Y., & Coxson, D. S. (2017). Externally held water – a key factor for hair lichens in boreal forest canopies. *Fungal Ecology*, *30*, 29–38. Available from https://doi.org/10.1016/j.funeco.2017.08.003.

Fang, L., Clausen, G., & Fanger, P. O. (1998). Impact of temperature and humidity on the perception of indoor air quality. *Indoor Air*, 80–90. Available from https://doi.org/10.1111/j.1600-0668.1998.t01-2-00003.x.

Fang, X., Winkler, J., & Christensen, D. (2011). Using EnergyPlus to perform dehumidification analysis on building America homes, In*HVAC and R Research*, *17*(3), 268–283. Available from https://doi.org/10.1080/10789669.2011.564260.

Gauslaa, Y. (2014). Rain, dew, and humid air as drivers of morphology, function and spatial distribution in epiphytic lichens. *Lichenologist (London, England)*, *46*(1), 1–16. Available from https://doi.org/10.1017/S0024282913000753.

Ge, H., Yang, X., Fazio, P., & Rao, J. (2014). Influence of moisture load profiles on moisture buffering potential and moisture residuals of three groups of hygroscopic materials. *Building and Environment*, *81*, 162–171. Available from https://doi.org/10.1016/j.buildenv.2014.06.021.

Gianangeli, A., Di Giuseppe, E., & D'Orazio, M. (2017). *Design and performance assessment of building counter-walls integrating Moisture Buffering \active\ devices, Energy Procedia* (132, pp. 105–110). Elsevier Ltd. Available from https://doi.org/10.1016/j.egypro.2017.09.652.

Greenspan, L. (1977). Humidity fixed points of binary saturated aqueous solutions. *Journal of Research of the National Bureau of Standards, Section A, Physics and Chemistry*, *81*(1), 89–96. Available from https://doi.org/10.6028/jres.081A.011.

Hameury, S. (2005). Moisture buffering capacity of heavy timber structures directly exposed to an indoor climate: a numerical study. *Building and Environment*, *40*(10), 1400–1412. Available from https://doi.org/10.1016/j.buildenv.2004.10.017.

Henderson, H.I., Parker, D., & Huang, Y.J. (2000). Improving DOE-2's RESYS routine: User defined functions to provide more accurate part load energy use and humidity predictions.

Holstov, A., Morris, P., Farmer, G., & Bridgens, B. (2015). Towards sustainable adaptive building skins with embedded hygromorphic responsiveness.

Huang, J., & Qian, X. (2008). Comparison of test methods for measuring water vapor permeability of fabrics. *Textile Research Journal*, *78*(4), 342–352. Available from https://doi.org/10.1177/0040517508090494.

Ingersoll, L. R. (1913). Indoor humidity [5]. *Science (New York, N.Y.)*, *37*(953), 524–525. Available from https://doi.org/10.1126/science.37.953.524-a.

ISO 24353. (2008). Hygrothermal performance of Building Materials and Products - determination of Moisture Adsorption/Desorption Properties in Response to Humidity Variation.

Janssen, H., & Roels, S. (2007). The hygric inertia of building zones: characterisation and application. IEA Annex.

Janssen, H., & Roels, S. (2009). Qualitative and quantitative assessment of interior moisture buffering by enclosures. *Energy and Buildings*, *41*(4), 382–394. Available from https://doi.org/10.1016/j.enbuild.2008.11.007.

Janz, M., & Johannesson, B. F. (2001). Measurement of the moisture storage capacity using sorption balance and pressure extractors. *Journal of Thermal Envelope and Building Science*, *24*(4), 316–334. Available from https://doi.org/10.1106/VRU2-LNV1-ME9X-8KKX.

JIS A1470-1. (2002). Test method of Adsorption/Desorption Efficiency For Building Materials to Regulate an Indoor Humidity - part 1: Response Method of Humidity.

Kerestecioglu, A., Swami, M., & Kamel, A. (1990). Theoretical and computational investigation of simultaneous heat and moisture transfer in buildings: effective penetration depth theory. *ASHRAE Transactions*, *96*(1), 447–454.

Kraniotis, D., Nore, K., Brückner, C., & Nyrud, A. Q. (2016). Thermography measurements and latent heat documentation of Norwegian spruce (Picea abies) exposed to dynamic indoor climate. *Journal of Wood Science*, *62*(2), 203–209. Available from https://doi.org/10.1007/s10086-015-1528-1.

Kreiger, B. K., & Srubar, W. V. (2019). Moisture buffering in buildings: a review of experimental and numerical methods. *Energy and Buildings*, *202*. Available from https://doi.org/10.1016/j.enbuild.2019.109394.

Kuenzel, H. (1965). Die Feuchtigkeitsabsorption von Innenoberflächen und Inneneinrichtungen. *Berichte Aus Der Bauforschung*, 101–116.

Kurnitski, J., Kalamees, T., Palonen, J., Eskola, L., & Seppänen, O. (2007). Potential effects of permeable and hygroscopic lightweight structures on thermal comfort and perceived IAQ in a cold climate. *Indoor Air*, *17*(1), 37–49. Available from https://doi.org/10.1111/j.1600-0668.2006.00447.x.

Latif, E., Ciupala, M. A., Tucker, S., Wijeyesekera, D. C., & Newport, D. J. (2015). Hygrothermal performance of wood-hemp insulation in timber frame wall panels with and without a vapour barrier. *Building and Environment*, *92*, 122–134. Available from https://doi.org/10.1016/j.buildenv.2015.04.025.

Latif, E., Lawrence, M., Shea, A., & Walker, P. (2015). Moisture buffer potential of experimental wall assemblies incorporating formulated hemp-lime. *Building and Environment*, *93*(2), 199–209. Available from https://doi.org/10.1016/j.buildenv.2015.07.011.

Latif, E., Lawrence, R. M. H., Shea, A. D., & Walker, P. (2018). An experimental investigation into the comparative hygrothermal performance of wall panels incorporating wood fibre, mineral wool and hemp-lime. *Energy and Buildings*, *165*, 76–91. Available from https://doi.org/10.1016/j.enbuild.2018.01.028.

Li, Y., Fazio, P., & Rao, J. (2012). An investigation of moisture buffering performance of wood paneling at room level and its buffering effect on a test room. *Building and Environment*, *47*(1), 205–216. Available from https://doi.org/10.1016/j.buildenv.2011.07.021.

Liuzzi, S., Hall, M. R., Stefanizzi, P., & Casey, S. P. (2013). Hygrothermal behaviour and relative humidity buffering of unfired and hydrated lime-stabilised clay composites in a Mediterranean climate. *Building and Environment*, *61*, 82–92. Available from https://doi.org/10.1016/j.buildenv.2012.12.006.

Loonen, R. C. G. M., Trčka, M., Cóstola, D., & Hensen, J. L. M. (2013). Climate adaptive building shells: State-of-the-art and future challenges. *Renewable and Sustainable Energy Reviews*, *25*, 483–493. Available from https://doi.org/10.1016/j.rser.2013.04.016.

Maskell, D., Thomson, A., Walker, P., & Lemke, M. (2018). Determination of optimal plaster thickness for moisture buffering of indoor air. *Building and Environment*, *130*, 143–150. Available from https://doi.org/10.1016/j.buildenv.2017.11.045.

McCullough, E. A., Kwon, M., & Shim, H. (2003). A comparison of standard methods for measuring water vapour permeability of fabrics. *Measurement Science and Technology*, *14*(8).

McGregor, F., Heath, A., Fodde, E., & Shea, A. (2014). Conditions affecting the moisture buffering measurement performed on compressed earth blocks. *Building and Environment*, *75*, 11–18. Available from https://doi.org/10.1016/j.buildenv.2014.01.009.

Moon, H. J., Ryu, S. H., & Kim, J. T. (2014). The effect of moisture transportation on energy efficiency and IAQ in residential buildings. *Energy and Buildings*, *75*, 439–446. Available from https://doi.org/10.1016/j.enbuild.2014.02.039.

Nguyen, D. M., Grillet, A. C., Bui, Q. B., Diep, T. M. H., & Woloszyn, M. (2018). Building bio-insulation materials based on bamboo powder and bio-binders. *Construction and Building Materials*, *186*, 686–698. Available from https://doi.org/10.1016/j.conbuildmat.2018.07.153.

Nore, K., Nyrud, A. Q., Kraniotis, D., Skulberg, K. R., Englund, F., & Aurlien, T. (2017). Moisture buffering, energy potential, and volatile organic compound emissions of wood exposed to indoor environments. *Science and Technology for the Built Environment*, *23* (3), 512–521. Available from https://doi.org/10.1080/23744731.2017.1288503.

Ojanen, T., & Salonvaara, M. (2003). A method to determine the moisture buffering effect of structures during diurnal cycles of indoor air moisture loads. In the 2nd International Conference on Building Physics, IBPC.

Osanyintola, O. F., & Simonson, C. J. (2006). Moisture buffering capacity of hygroscopic building materials: Experimental facilities and energy impact. *Energy and Buildings*, *38* (10), 1270–1282. Available from https://doi.org/10.1016/j.enbuild.2006.03.026.

Osanyintola, O. F., Talukdar, P., & Simonson, C. J. (2006). Effect of initial conditions, boundary conditions and thickness on the moisture buffering capacity of spruce plywood. *Energy and Buildings*, *38*(10), 1283–1292. Available from https://doi.org/10.1016/j.enbuild.2006.03.024.

Ouméziane, Y.A., Bart, M., Moissette, S., Lanos, C., Prétot, S., & Collet, F. (2011). Hygrothermal behaviour of a hemp concrete wall: influence of sorption isotherm modelling.

Padfield, T. (1998). The role of absorbent building materials in moderating changes of relative humidity.

Palumbo, M., Lacasta, A. M., Holcroft, N., Shea, A., & Walker, P. (2016). Determination of hygrothermal parameters of experimental and commercial bio-based insulation materials. *Construction and Building Materials*, *124*, 269–275. Available from https://doi.org/10.1016/j.conbuildmat.2016.07.106.

Pavlík, Z., & Černý, R. (2008). Experimental assessment of hygrothermal performance of an interior thermal insulation system using a laboratory technique simulating on-site conditions. *Energy and Buildings*, *40*(5), 673–678. Available from https://doi.org/10.1016/j.enbuild.2007.04.019.

Pavlík, Z., & Černý, R. (2009). Hygrothermal performance study of an innovative interior thermal insulation system. *Applied Thermal Engineering*, *29*(10), 1941–1946. Available from https://doi.org/10.1016/j.applthermaleng.2008.09.013.

Rahim, M., Douzane, O., Tran Le, A. D., Promis, G., Laidoudi, B., Crigny, A., ... Langlet, T. (2015). Characterization of flax lime and hemp lime concretes: hygric properties and moisture buffer capacity. *Energy and Buildings*, *88*, 91–99. Available from https://doi.org/10.1016/j.enbuild.2014.11.043.

Ramos, N.M., & De Freitas, V.P. (2006). Evaluation strategy of finishing materials contribution to the hygroscopic inertia of a room. In Proceedings of the 3rd International Building Physics Conference - Research in Building Physics and Building Engineering (pp. 543–548).

Ramos, N. M. M., & De Freitas, V. P. (2009). An experimental device for the measurement of hygroscopic inertia influence on RH variation. *Journal of Building Physics*, *33*(2), 157–170. Available from https://doi.org/10.1177/1744259109104885.

Reichert, S., Menges, A., & Correa, D. (2015). Meteorosensitive architecture: biomimetic building skins based on materially embedded and hygroscopically enabled responsiveness. *CAD Computer Aided Design*, *60*, 50–69. Available from https://doi.org/10.1016/j.cad.2014.02.010.

Reinikainen, L. M., & Jaakkola, J. J. K. (2003). Significance of humidity and temperature on skin and upper airway symptoms. *Indoor Air*, *13*(4), 344–352. Available from https://doi.org/10.1111/j.1600-0668.2003.00155.x.

Rode, C., Holm, A., & Padfield, T. (2004). A review of humidity buffering in the interior spaces. *Journal of Thermal Envelope and Building Science*, *27*(3), 221–226. Available from https://doi.org/10.1177/1097196304040543.

Rode, C., Peuhkuri, R.H., Hansen, K.K., Time, B., Svennberg, K., Arfvidsson, J., & Ojanen, T. (2005). NORDTEST project on moisture buffer value of materials. Paper presented at the AIVC Conference 'Energy performance regulation': Ventilation in relation to the energy performance of buildings.

Rode, C., Peuhkuri, R.H., Mortensen, L.H., Hansen, K.K., Time, B., Gustavsen, A., & Arfvidsson. (2005). Moisture buffering of building materials.

Roels, S., & Janssen, H. (2006). A comparison of the Nordtest and Japanese test methods for the moisture buffering performance of building materials. *Journal of Building Physics*, *30*(2), 137–161. Available from https://doi.org/10.1177/1744259106068101.

Salonvaara, M., Ojanen, T., Holm, Künzel, H.M., & Karagiozis, A.N. (2004). Moisture buffering effects on indoor air quality-experimental and simulation results.

Simonson, C. J., Salonvaara, M., & Ojanen, T. (2004). Heat and mass transfer between indoor air and a permeable and hygroscopic building envelope: part I—field measurements. *Journal of Thermal Envelope and Building Science*, *28*(1), 63–101. Available from https://doi.org/10.1177/1097196304044395.

Stazi, F., Tittarelli, F., Politi, G., Di Perna, C., & Munafò, P. (2014). Assessment of the actual hygrothermal performance of glass mineral wool insulation applied 25 years ago in masonry cavity walls. *Energy and Buildings*, *68*, 292–304. Available from https://doi.org/10.1016/j.enbuild.2013.09.032.

Svennberg, K., Lengsfeld, K., Harderup, L. E., & Holm, A. (2007). Previous experimental studies and field measurements on moisture buffering by indoor surface materials. *Journal of Building Physics*, *30*(3), 261–274. Available from https://doi.org/10.1177/1744259107073221.

Toman, J., Vimmrová, A., & Černý, R. (2009). Long-term on-site assessment of hygrothermal performance of interior thermal insulation system without water vapour barrier.

Energy and Buildings, *41*(1), 51–55. Available from https://doi.org/10.1016/j.enbuild.2008.07.007.

Tran Le, A. D., Maalouf, C., Douzane, O., Promis, G., Mai, T. H., & Langlet, T. (2016). Impact of combined moisture buffering capacity of a hemp concrete building envelope and interior objects on the hygrothermal performance in a room. *Journal of Building Performance Simulation*, *9*(6), 589–605. Available from https://doi.org/10.1080/19401493.2016.1160434.

Tran Le, A. D., Maalouf, C., Mai, T. H., Wurtz, E., & Collet, F. (2010). Transient hygrothermal behaviour of a hemp concrete building envelope. *Energy and Buildings*, *42*(10), 1797–1806. Available from https://doi.org/10.1016/j.enbuild.2010.05.016.

Vallati, O., Brambilla, A., Spalletti, G., Frigerio, G., & Ranzi, G. (2018). An initial study for the developement of an adaptive timber device for moisture buffering applications. In COST Action TU1403: Adaptive Facades Network Final Conference: Facade 2018 - Adaptive.

Vereecken, E., Roels, S., & Janssen, H. (2011). In situ determination of the moisture buffer potential of room enclosures. *Journal of Building Physics*, *34*(3), 223–246. Available from https://doi.org/10.1177/1744259109358268.

Woloszyn, M., Kalamees, T., Abadie, M. O., Steeman, M., & Sasic Kalagasidis, A. (2009). The effect of combining a relative-humidity-sensitive ventilation system with the moisture-buffering capacity of materials on indoor climate and energy efficiency of buildings. *Building and Environment*, *44*(3), 515–524. Available from https://doi.org/10.1016/j.buildenv.2008.04.017.

Woods, J., Winkler, J., & Christensen, D. (2013). Evaluation of the effective moisture penetration depth model for estimating moisture buffering in buildings. *Contract (New York, N.Y.: 1960)*, *303*, 275–3000.

Wu, Y., Gong, G., Yu, C. W., & Huang, Z. (2015). Proposing ultimate moisture buffering value (UMBV) for characterization of composite porous mortars. *Construction and Building Materials*, *82*, 81–88. Available from https://doi.org/10.1016/j.conbuildmat.2015.02.058.

Yang, J., Fu, H., & Qin, M. (2015). Evaluation of Different Thermal Models in EnergyPlus for Calculating Moisture Effects on Building Energy Consumption in Different Climate Conditions. In Procedia Engineering (Vol. 121, pp. 1635–1641). Elsevier Ltd. https://doi.org/10.1016/j.proeng.2015.09.194

Yang, X., Vera, S., Rao, J., Ge, H., & Fazio, P. (2007). Full-scale experimental investigation of moisture buffering effect and indoor moisture distribution. Thermal Performance of Exterior Envelopes of Whole Buildings X.

Zhang, H., Yoshino, H., & Hasegawa, K. (2012). Assessing the moisture buffering performance of hygroscopic material by using experimental method. *Building and Environment*, *48*(1), 27–34. Available from https://doi.org/10.1016/j.buildenv.2011.08.012.

Zhang, M., Qin, M., Rode, C., & Chen, Z. (2017). Moisture buffering phenomenon and its impact on building energy consumption. *Applied Thermal Engineering*, *124*, 337–345. Available from https://doi.org/10.1016/j.applthermaleng.2017.05.173.

Hygrothermal modelling

6

Undesired moisture within the building envelope can cause serious damage to building materials, reduce the service life of the whole building, and present a health hazard for the occupants. Yet, increasingly stringent energy efficiency provisions are driving a move towards airtight envelopes that minimise moisture exchange between indoors and outdoors, which makes it more difficult to mitigate moisture-related problems. This indicates the need for a careful envelope design that takes account of hygrothermal processes from the early design phase to ensure resilience in the case of possible failures.

General knowledge about the phenomenon and the response of different build-ups is no longer sufficient to guarantee a successful outcome by reducing or completely eliminating the risk of hygrothermal issues. Technological advances in materials composition, building components and construction processes have meant that the physical response of a building to moisture is less predictable, increasing the need for accurate building physics models. Several hygrothermal models can be used to describe the mono- or multi-dimensional hygrothermal performance of a building envelope or a whole building (Delgado, Ramos, Barreira, & Freitas, 2010). However, these models rely on differential equations that are difficult to solve without specific computational software. For this reason, several computer-based tools have recently been developed to help designers predict the hygrothermal behaviour of buildings. The main difference between them is their degree of complexity for the user. This reflects how or whether the following parameters are included: moisture exchange dimensions (mono- or multi-dimensional), type of assessment (transient vs steady-state), quality of the assumptions, and the availability of various types data, such as material properties, weather and construction (Delgado, Barreira, Ramos, & Freitas, 2012).

The intended use of a particular software may also influence the complexity of its application, the time required for the computation, the inputs needed to perform the assessment, the final results and the accuracy of the findings. It is sometimes difficult for designers and researchers interested in hygrothermal assessment to navigate through the available types of software to identify the one that is best suited to their purpose. This chapter analyses some of the most widely used hygrothermal simulation software in relation to their applicability, required inputs and assumptions, and ease of use. The aim is to explain the potential applications and the criticalities embedded in current simulation tools. For this reason, the specific mathematical model used in each tool is not discussed in detail. Rather, the focus is on the overall assumptions and inputs that underpin the accuracy and sophistication of the software.

Moisture and Buildings. DOI: https://doi.org/10.1016/B978-0-12-821097-0.00003-5

6.1 Hygrothermal modelling and assessment

Moisture-related building defects and pathologies are very difficult to predict, yet their economic and health implications make the topic of high importance during the building design and maintenance phases. Generally, building regulations allow hygrothermal performance to be assessed using either a steady-state or transient approach. Considering the number of tools and software available, as well as the complexity of the topic, the ASTM Standard (ASTM, 2016) has developed a framework for transient hygrothermal assessments that aims to standardise the calculations, provide freedom in the selection of tools or software, and ensure a consistent approach to moisture control across different projects. This basic principle is embedded in the process, which follows the same steps regardless of the assumptions, calculation choices and tools employed.

These steps are described in Fig. 6.1:

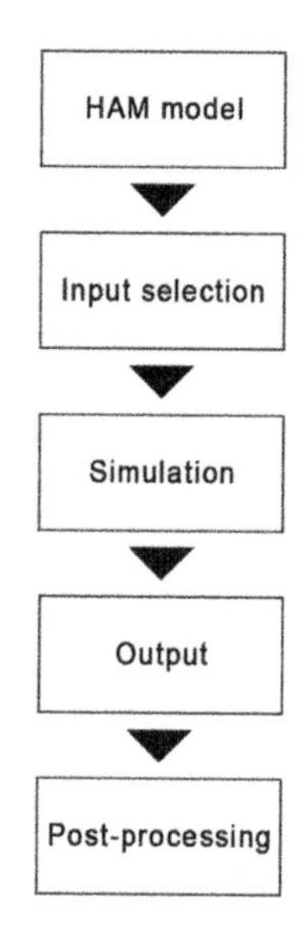

Figure 6.1 Hygrothermal modelling steps.

1. Definition of the heat, air and moisture (HAM) model and type of assessment. This involves identifying the model functionalities necessary for the assessment, that is, the set of functions and assumptions needed in relation to modelling, simulation process and analytical capabilities. From this first step, it is possible to identify the appropriate tool or software to provide the functions required.
2. Selection of input and assumptions. This involves defining those properties and data necessary to contextualise the analysis, including geometry of the envelope, materials properties, internal and external climates and boundary conditions.
3. Simulation or calculation through the selected tool. This is the core of the assessment, where inputs are used to run the calculation, solve the mathematical equations and provide the numerical outputs.

4. Collection of the output. The raw data resulting from the calculation include, but are not limited to, the hourly values of temperature, heat flux, water content, relative humidity and moisture distribution and evolution.
5. Post-processing. Analysis of the raw data according to the purpose of the analysis usually refers to a standard or building code. It can be performed by the software or tools as an additional step or can be done manually.

6.1.1 Heat, air and moisture model

6.1.1.1 Temporality

Steady-states methods are usually mono-dimensional. They consider vapour diffusion through building materials and are used to quantify the maximum amount of accumulated moisture within a building component. These methods are simple and easy to apply, requiring only general information about the materials and climates, such as monthly averages of temperature and relative humidity. Thus the risk of condensation can be calculated using a few general tools as a normal calculation spreadsheet. However, these methods must be considered as preliminary and generic assessment tools. Although steady-state assessment may be suitable for a quick comparison between different building components, transient assessments are more reliable and robust for accurate prediction over the building lifespan. Transient methods can account for dynamic changes and cyclical conditions that a building may encounter during its life. In particular, transient assessments acknowledge drying-out processes and the dependence of the material properties on temperature and moisture content, liquid transfer, air movement within the envelope, hygroscopic capacity of materials, and variable boundary conditions. All these parameters strongly influence the hygrothermal performance of a building and cannot be overlooked when assessing the risk of moisture-related damage. Transient assessments rely on combined HAM modelling (as described in Chapter 2, Principles of hygrothermal processes), which can be performed using specific hygrothermal simulation software. The difference between these tools lies in their mathematical sophistication and the degree of complexity with which the boundary conditions are described. (Straube & Burnett, 2001).

Different software uses different mathematical models to describe heat and mass transfer through the building envelope. Although knowledge of the mathematical formulation is not an essential requirement for applying the software, the user must be aware of the phenomena that need to be included and the types of boundary conditions that should be considered. For example, most models do not consider air transport or rain wetting, making them less than optimal in tropical contexts where monsoons drive the building's hygrothermal performance. Further, as seen in Chapter 2, hygrothermal models can be based on different equations, which differ in relation to the assumed driving force. Unlike heat transfer, there is still no agreement about the main driver of moisture processes. Moisture content, (Luikov, 1964; Philip & De Vries, 1957) relative humidity and porous radius (Hens, 1996), moisture content and water vapour pressure (Salonvaara, Zhang, & Karagiozis, 2011), relative humidity (Künzel, 1995) and suction and water vapour pressure (Burch & Chi, 1997) are all considered possible main drivers. It is therefore important to understand which type of transport processes are considered, and what is assumed to be the driver for moisture transfer.

6.1.1.2 Dimensionality

The available tools can be classified according to the HAM dimension that is considered: 1D, which considers homogenous wall cross sections, 2D, which accounts for anisotropic material properties or articulated wall build-ups (e.g. localised presence of structural members, intersection between walls and roof) or 3D, for whole building assessments. Mono-dimensional models are used to assess the hygrothermal behaviour of the building envelope and consider the HAM transfer perpendicular to the component in the analysis. Bi-dimensional models account for the HAM transfer in two directions, and are used to evaluate hygrothermal performance at particular intersections between different building components, such as a wall and a floor, or in the presence of discontinuity in the layers, such as in proximity to thermal bridges. Tri-dimensional models are usually referred to as whole building models, and are based on the integration of thermal transient analysis with the other two types of models. This type of assessment is based on the need to describe the internal climate with more accuracy: many of the 1D and 2D tools require the user to input the internal climate but, in reality, the indoor air hygrothermal conditions depend on the envelope. Thus, the envelope hygrothermal performance and the indoor climate cannot be considered as two separate entities. Whole building HAM simulation tools account for these interdependencies.

> *"The hygrothermal balances consider the normal flows of heat by conduction, convection, and radiation; moisture flows by vapour diffusion, convection, and liquid transport and airflows driven by natural, external, or mechanical forces. However, in whole building heat, air and moisture analyses, combined forms which involve advection heat flows and conversion of latent heat should also be considered. Likewise, the thermal conditions influence very strongly the moisture conditions and natural airflows"*
>
> *Woloszyn & Rode (2008)*

However, greater computational capacity also means longer calculation time, and it is therefore important to identify the specific assessment needs and select the best tool accordingly. In order to do so, it is necessary to understand how these pieces of software work and what their potentialities are.

6.1.1.3 Numerical method-solver

Another difference among the software is the solver used. There are different numerical methods for solving the differential equation governing the HAM transfer, including the following (Delgado et al., 2012):

- finite difference methods (FDM), which approximate the derivatives to differential quotients and calculate the approximation of the solution in time or space points; in other words, they are discretisation methods;

- finite control volume method (FCV), which is usually used in fluid dynamics or applied to physical conservation laws. It discretises the space where the variables live into controlled volumes and evaluates the flux at the boundary between volume in a conservative way;
- finite element methods (FEM), which divide the domain of the solution into a discrete number of non-overlapping elements where the functions are approximated to local functions (such as polynomials), resulting in a system of equations that describe the whole problem;
- transfer function method, which describes the systems through input-output relations.

The numerical method used influences the accuracy of the results obtained and the computational time.

6.1.1.4 Time step

Transient hygrothermal assessments provide the ability to deal with dynamic time-dependent phenomena, such as daily, monthly and seasonal climatic variations. While steady state assessments use averaged conditions, transient tools allow hygrothermal performance to be considered on a much finer time grid, acknowledging the effects of cyclical conditions and changing vapour pressure on the building elements, as well as storage (or buffering) and wetting-and-drying processes. Most of the simulation software uses hourly time steps, as the majority of weather files are expressed with this interval. Hourly time steps also represent an acceptable compromise between accuracy and manageability – a one-hour climatic value is a good approximation of more refined data, such as 15-minute intervals, and creates a yearly series with only 8760 values, a number that is easy to post-process using a normal calculation sheet.

It is important to note that most software programs require a one-year series, which is repeated several times to obtain longer and more accurate exposure. However, this approach overlooks the importance of multi-year phenomena. Building components and materials change their behaviour over the years due to natural processes such as ageing, or long-term damage such as corrosion or biodeterioration. The currently available hygrothermal tools do not allow these types of degradation to be modelled in a multi-year analysis (Chung, Wen, & Lo, 2019). This affects the overall potential of hygrothermal software, limiting its scope to a certain extent. However, software remain powerful tools that can help designers to deliver healthy indoor environments and durable constructions.

6.1.2 Input

Hygrothermal modelling assists in the assessment of energy, moisture and air balances. The hygrothermal performance of a building or building component depends on the HAM transfer between the indoors, the outdoors and the building envelope. Thus, in order to perform the assessment, it is necessary to set several assumptions and inputs in regard to build-up, materials properties and climate. These assumptions and inputs can be categorised into four groups (Fig. 6.2):

- Information related to the HAM model. This group includes all the mathematical references that the software uses for the hygrothermal assessment. These assumptions are not

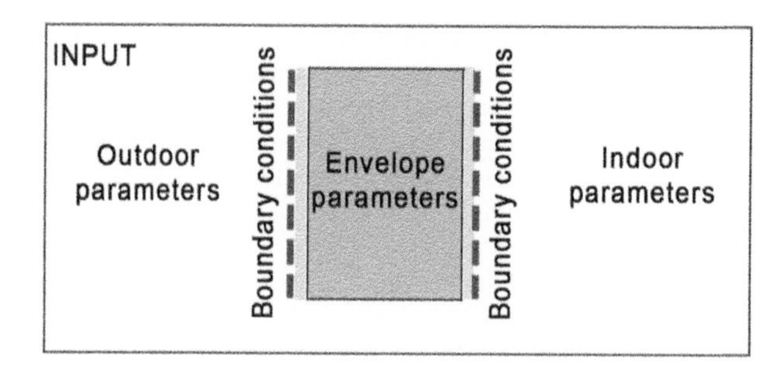

Figure 6.2 Different types of input for hygrothermal modelling.

directly controllable by the user, but the selection of the right software can provide good control over the model to be used.
- Information related to the indoors. This comprises all those parameters that characterise the indoor climate and the boundary between the envelope and the interior air.
- Information related to the outdoors. This comprises all those parameters that characterise the outdoor climate and the boundary between the envelope and the exterior context.
- Information related to the envelope. This comprises all those parameters that can be used to characterise the envelope, including the geometry, materials, thickness, presence of ventilation layers or air infiltration, and materials properties.

Hygrothermal simulations require all these parameters in order to compute and assess the HAM transfer. However, the various software packages differ in terms of their accuracy and the range of choice of parameters they allow. Some, for example, include a database from which different pre-set averaged values can be selected, while others offer a broader and customisable input selection. It is usually the case that more accurate and sophisticated tools need more information, which can be hard to generalise and might need to be defined through empirical experiments. It is an open question as to when the simulation can be considered accurate enough in regard to the design stage at which it is needed.

6.1.2.1 Boundary conditions at the surface

The boundary conditions are the set of assumptions that are used to describe the interfaces between two different elements. The most important are those that concern the interfacial flow between the climate, either internal or external, and the building component, as this determines the extent to which one influences the other. To perform hygrothermal assessments it is important to define those coefficients that influence these exchanges and regulate the HAM transfer between the outdoors, the envelope and the indoors. The hygrothermal software determines how these coefficients are accounted for and their accuracy. For example, some use a combination of the convective and long-wave radiation coefficients, others rely on a temperature-dependent convection factor or view factors for the determination of long-wave radiation.

6.1.2.1.1 Convective heat transfer coefficient

Each surface exchanges heat with the adjacent thin layer of air through convection. This phenomenon can be expressed as a function of a specific coefficient (Hagentoft, 2001):

$$Q_c = h_c \cdot (T_{\text{air}} - T_{\text{surface}}) \tag{6.1}$$

However, this convective heat transfer coefficient (h_c) cannot be univocally defined. The reason lies in the complexity of the phenomena that determine the coefficient, which include multiple factors such as the characteristics of the surface, geometry, air flow direction and magnitude, among many others. On the interior side of the construction, convection is mainly driven by temperature differences, and the method employed to define this coefficient distinguishes between the type of surface under consideration. Different sources identify different functions that can be used to determine the coefficient, mainly as a function of the temperature difference between the air and the surface. More recently, the determination of the convective heat transfer coefficient has been simplified by introducing nominal values, as reported in Table 6.1. These values can differ significantly, and it is important to cite the reference used for the hygrothermal assessment in order to clarify the range of validity of the analysis.

In contrast, the external surface is more subject to the influence of wind; hence, the convective heat transfer coefficient is driven by wind direction and speed. Consensus is yet to be reached on the correlation between the elements. The hygrothermal simulation tools usually provide internal calculation of the coefficient or refer to the model that must be used for the calculation.

6.1.2.1.2 Radiative heat transfer coefficient

The building surfaces exchange heat with the other terrestrial surfaces through long-wave radiation transfer which, following the Stefan-Boltzmann Law, depends on the surrounding and terrestrial surface temperatures, the long-wave emissivity of the surface and the Stefan-Boltzmann constant. The radiative heat transfer coefficient (h_r) lumps all the parameters and allows the heat flux to be expressed as a linear function (Hagentoft, 2001):

$$Q_r = h_r \cdot (T_{\text{air}} - T_{\text{surface}}) \tag{6.2}$$

Table 6.1 Convective heat transfer coefficient according to different sources.

	EN (15026 2007)	**(Hens, 2017)**	**(Künzel, 1995)**
Vertical heat flux (walls)	2.5	3.5	in the range 3 to 10
Horizontal upward heat flux (ceilings)	5.0	5.5	in the range 3 to 10
Horizontal downward heat flux (floors)	0.7	1.2	in the range 3 to 10

where h_r is calculated as a function of the surface, surrounding surfaces and terrestrial temperatures. Although it clearly varies in time, due to fluctuations in temperature, in most cases it is assumed as a constant value that ranges between 3 and 6 W/m^2K (Delgado et al., 2012). However, to assume this coefficient as constant, it is necessary to separate the effects of the solar radiation on the surface temperature, due to its significant impact. To include this component, the models rely on another coefficient—short-wave radiation absorptivity. Solar radiative heat flux is defined as a function of the solar radiation normal to the surface and this coefficient. It is possible to compute this additional energy hitting the surface by specifying the geometry of the component (inclination and orientation) and the short-wave absorptivity of the finishing material (colour and brightness) in the analysis.

6.1.2.1.3 Incident rain on the façade

Rain is one of the main sources of water that hit a façade. Some models allow this component to be included in the calculation by requiring a coefficient that can indicate the amount of water that is allowed to deposit on the façade. However, this coefficient depends on a number of factors, such as building geometry, wind speed, rain intensity and wind direction (Delgado, Azevedo, & Guimarães, 2019); hence, it is characterised by great variability (Nore, Blocken, Jelle, Thue, & Carmeliet, 2007). Hygrothermal software computes this component by estimating the amount of water that can reach the surface during a rain event and determining the quantity that can be absorbed by the material. The first value depends on a rain coefficient and the weather file, specifically, rain intensity, wind velocity and wind direction (Nore et al., 2007). Different rain models assume different values for the coefficient, usually ranging between 0.02 and 0.26 s/m (Blocken, 2004). The quantity of water absorbed is instead determined by the precipitation absorptivity, which accounts for splash off.

6.1.2.2 Indoor climate

The indoor climate determines the hygrothermal pressure acting on the envelope. The inputs required for the HAM transfer computation are the temperature and relative humidity at the given time-step (usually 1 hour). Although it seems fairly straightforward, this step poses a significant risk in relation to the reliability of the results. Unless the hygrothermal simulation is performed on a previously constructed case study and it is possible to monitor the indoor climate, the definition of these variables at an hourly resolution during the design stage presents a real challenge. Any schedule that can be used introduces a significant source of error in the simulation, mainly due to the unpredictability of the building occupants – who may use the heating, ventilation and air conditioning (HVAC) differently from what was assumed or perform activities that generate high levels of moisture – and the interdependencies between the envelope hygrothermal behaviour and the indoor climate. Several standards can be used to generate the values, including (ISO EN 13788, 2012), (ISO EN 15026, 2007), or (Standard 160–2016. Criteria for Moisture-Control Design Analysis in Buildings, 2016).

6.1.2.3 Outdoor climate

The parameters that characterise the outdoor climate are more refined and numerous compared to those for the indoor environment. Temperature and humidity account for only a small part of the conditions at the exterior of the building that can influence hygrothermal exchange through the envelope. As discussed, hygrothermal modelling requires clear information in regard to the boundary conditions at the surface, which strictly depend on other climatic parameters such as wind, solar radiation and rain. The software requires all these parameters, specified at the right time-scale. Hygrothermal assessments are more reliable when computed for a number of years (e.g., ASHRAE 160 indicates 10 consecutive years), due to the cyclical process of wetting and drying which can lead to long-term issues. However, multi-year analysis is usually not allowed, so the HAM transfer is computed on the basis of a reference year analysis (Chung et al., 2019). The file that collects this information is usually called the Moisture Reference Year (MRY), which should be more severe than the average climate to err on the side of caution. The literature presents different statistical methods to generate this dataset, based on the statistical analysis of monitored meteorological data (Salonvaara et al., 2011). The selection of one over the other is usually dictated either by the standard followed for the analysis or by the designer.

The outdoor climatic parameters needed are:

- dry bulb temperature (°C)
- relative humidity (%)
- wind velocity (m/s) and direction – counted clockwise from north over east
- rain (mm)
- ambient air pressure (hPa)
- direct or global solar radiation on the horizontal plane (W/m^2)
- diffuse solar radiation on the horizontal plane (W/m^2)
- cloud index.

6.1.2.4 Envelope and materials

The envelope characterisation is a very delicate part of the hygrothermal assessment. Material selection and modelling should be carefully considered, as they can greatly influence the final hygrothermal response. The accuracy of the assessment depends on the accuracy of the input data, including the envelope modelling. Each type of simulation software prescribes a different approach to modelling and characterising the envelope, but most contain a built-in database from which it is possible to select common build-ups and construction materials.

6.1.2.4.1 Geometry

The geometry of the building component is one of the first analytic steps that must be undertaken in hygrothermal assessments. The type of information required depends on the software and the scope of the analysis. 1D analysis requires only the cross section, as it assesses only the mono-directional HAM transfer, while 2D analysis allows for the assessment of components' intersections and more refined geometries. In contrast, 3D assessments usually look at the whole building; hence, the resolution of the building component varies greatly from software to software, although the majority assess HAM transfer through the envelope as mono-directional. The basic principle for defining the geometry is to divide the building component into a series of consequent and adjunct layers of materials. Although this might be a good approximation, it completely overlooks the quality of construction. Most of the analyses are conducted on ideal building components, where the materials are all in perfect contact with one another. Gaps, cracks and discontinuities create breaks in the capillarity flow and influence the airflow and moisture transfer. However, software programs are rarely able to account for these imperfections, which introduces systematic errors into the results.

6.1.2.4.2 Materials

The materials properties required for a transient assessment are characterised by a higher resolution compared to those used in traditional steady-state assessments, where a single value for thermal conductivity and vapour permeability is necessary. Since the materials' response to hygrothermal stresses and their properties are also influenced by humidity and temperature, a transient approach acknowledges these interdependencies, and the properties of a material are expressed as curves or function of temperature and moisture content, rather than being defined by a single constant value. Unfortunately, these functions also vary by manufacturer, and two similar products may exhibit completely different behaviour; this type of information is only available for a limited number of products. Different sources of information can be used for the analysis, and the quality of this information determines the quality and accuracy of the final assessment. The following classification is based on the source (ASTM, 2016):

1. Full functional properties. The material properties are precisely known, having been obtained from the specific manufacturer or tested on samples, and information on hygroscopic and capillary regime is available in full;
2. Literature. Complete material properties data is available in literature that provides documentation on the source and testing results.
3. Approximated data. Generic properties data is used, the sample origin is not defined and multiple origins may be used.

However, most software includes a built-in materials database, based on the empirical characterisation of commercial products. Generally, the following information is needed for each material:

- bulk density
- porosity
- specific heat capacity
- thermal conductivity
- water vapour permeability
- water absorption coefficient
- moisture storage functions.

6.1.3 Simulation, output and post-processing

Hygrothermal simulations solve the transfer equations using an iterative calculation process. The number of convergence failures that occur during the simulation is a general indicator of the quality of the calculation process. If no failure occurs, this indicates that the calculation has not encountered any numerical problems. However, failures are not a cause for concern when the change in total water content during the calculation is equal, or very similar, to the sum of the moisture fluxes across the surfaces. Hence, these are the first parameters to check to assess the reliability of the results.

Most transient hygrothermal tools generate output in the form of a series of hourly values, such as temperature, heat flux, water content, relative humidity, and moisture distribution and evolution. These series may be used to validate the results using real measured data or results from similar analyses; this process provides an additional level of robustness and reliability to the assessment. Usually, the raw data are also used for the reporting phase, which can be prescribed and standardised by the specific building standard or code used for the assessment. However, raw data are not usually used for specific risk assessment but require a post-processing phase. This analysis can be performed by the software or tools as an additional step, or manually.

6.1.4 Hygrothermal modelling and assessment process

Fig. 6.3 summarises the hygrothermal assessment process and the relevant decisions and assumptions that need to be made.

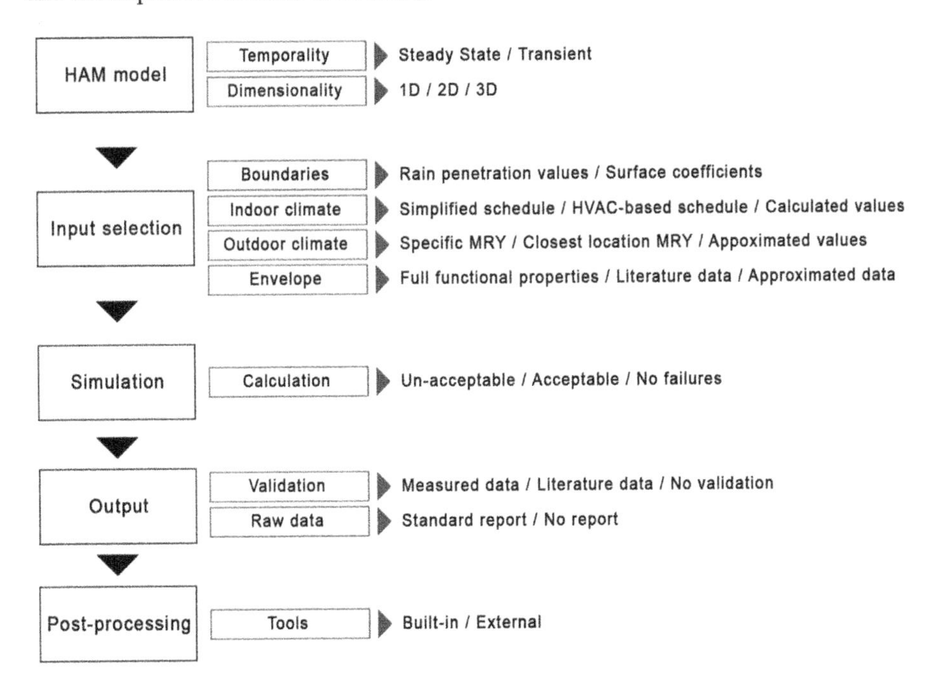

Figure 6.3 Hygrothermal modelling and assessment process.

6.2 Transient hygrothermal simulation software

The various types of simulation software account for the above parameters and assumptions in different ways. The accuracy of the results is heavily dependent on the accuracy of the assumptions and the resolution of the calculation. The following sections describe the features and assumptions of several hygrothermal software programs. This is not intended to be a complete database, but it includes the most widely used software currently available. The differences between them are examined in relation to their implications for designers or other users seeking to assess the hygrothermal behaviour of a building or a building component during the design stage. The mathematical formulations can be found in other resources with a specific focus on this topic (Delgado et al., 2012).

6.2.1 Applications

Transient hygrothermal software usually runs hourly-based simulations, expressing the outputs as hourly values. For 1D or 2D envelope assessments, the minimum outputs are surface temperature, relative humidity and water content at a given point; for whole building simulations, it is possible to obtain indoor temperature and relative humidity. The software usually gives back a raw result file that contains this information, which can then be analysed using external tools or built-in post-processors. Depending on the application, different kinds of analysis can be performed, including:

1D:
- Design and assessment of building components (as a substitute for steady-state methods, such as Glaser)
- Investigation of possible moisture-related damage
- Mould growth and assessment
- Development or improvement of building material products
- Design and evaluation of insulation strategies
- Evaluation of ventilated systems
- Drying problems of the envelope

2D:
- Calculation of thermal bridges and evaluation of moisture-based issue
- Design and evaluation of solutions for thermal bridges
- Drying problems of the envelope solution
- Development and assessment of technological systems

Whole-building:
- Transient calculation of energy demand with moisture-dependent values
- Assessment of indoor comfort accounting for moisture buffering.

6.2.2 Commercial hygrothermal tools

This section reviews some of the most widely used commercial software. The aim is not to describe the individual mathematical models but to present a snapshot of the software's features, advantages and drawbacks. This provides designers, researchers or general users with a framework of currently available tools for conducting envelope or whole building hygrothermal assessments.

6.2.2.1 WUFI

WUFI is a family of hygrothermal software developed by the Fraunhofer Institute in Building Physics. It is considered to have the greatest hygrothermal capabilities of all the available software (Pallin, Boudreaux, Shrestha, New, & Adams, 2017) and its ease of use has made it one of the most popular programs in the market. The family includes different tools to cover a great variety of purposes, such as: WUFI-Pro for 1D assessments; WUFI-2D, which accounts for bi-dimensional HAM flux; and WUFI-Plus for whole building simulations. There are also several free post-processors that can be used to analyse the data and which have been designed to facilitate the application of hygrothermal assessment to real case scenarios. These add-ons include a mould growth risk calculator, based on the VTT or biohygrothermal model, and a corrosion evaluation tool.

The envelope assessment is based on a validated model that allows for realistic calculation of the hygrothermal performance of a multi-layer building component. Heat transfer is considered in relation to conduction, enthalpy flows, and short-wave and long-wave radiative components, while moisture transfer is considered in relation to vapour diffusion, capillary liquid transfer and liquid surface diffusion. The whole building tool combines the benefits of envelope analysis with the features that characterise thermal transient assessments. The model accounts for eventual moisture sources and sinks within the room, moisture exchange with the envelope through capillarity, diffusion and sorption-desorption phenomena, and standard indoor and outdoor thermal exchanges.

The main advantage of WUFI is its user-friendly interface, which is built as a series of drop-down menus and clickable buttons that make it easy for non-experts to use. The software includes an extensive built-in database for climate, materials and other parameters, but also offers a high degree of customisation and flexibility to suit a variety of purposes. This tool represents a good balance between the accuracy of the results and the usability of the software. However, studies have identified two major limitations of WUFI. First, the materials properties are not expressed as a function of the temperature, which has been shown to have a significant impact on the behaviour of organic materials, such as wood (Pallin et al., 2017). Second, the software neglects air infiltrations that may occur through the envelope. Although it allows for the air-leakage rate to be defined manually, this relies on users' knowledge of how to assess this rate. Air infiltration usually depends on the permeability of the envelope and the pressure gradient, and its impact on the durability of building materials indicates the urgent need for current hygrothermal tools to be updated to include a more reliable modelling technique (Prowler & Trechsel, 2008). It must be said that this shortfall is common to almost all simulation software, so it is not repeated for each of the other packages discussed below. However, it remains a highly significant research gap that must be addressed (Table 6.2).

Table 6.2 WUFI simulation tool.

	WUFI
Dimension	1D - 2D-whole building
HAM model: moisture driving potential	Vapour pressure and moisture content
Moisture transfer	Vapour diffusion, liquid transport
Numerical method	Finite Control Volume
Boundary conditions at the surface	Combined constant coefficient for long-wave radiation and convection
Wind-driven rain	Yes
Materials properties	Function of relative humidity

Expertise required: Depending on the tool used, a general understanding of the building envelope is necessary, but the software interface allows the user to learn quickly and provides a step-by-step tutorial for non-experts.

Target user: Designers, engineers, building product and construction companies, researchers.

Reference: Künzel (1995).

6.2.2.2 DELPHIN

DELPHIN is a software for 1D or 2D hygrothermal assessment of building components, developed by the Institute of Building Climatology of the Technical University of Dresden. It includes a model of transport of heat, air, moisture, pollutants and salt within porous materials or multi-layer structures. DELPHIN accounts for moisture transport by vapour diffusion and liquid transport, but is also one of the few tools to consider moist air movements within the envelope. This software can take account of gravitation, radiative heat transfer (long-wave, shortwave), wind-driven rain, air pressure and air infiltration, which is calculated based on wind pressure. The materials properties depend on temperature and consider eventual 3D anisotropy.

One of the benefits of this software is the ability for users to calculate the indoor climate from a simplified room used as the boundary conditions for the envelope assessment (Woloszyn & Rode, 2008). Further, the integrated post-processing tools for mould calculation (VTT and isopleths), the direct control over the inputs and the simple graphical interface make DELPHIN one of the most complete and easy to use software packages currently available. One of the main limitations of this software is its inability to describe all the material properties as temperature-dependent, thus neglecting the potential impact of temperature on timber and timber-based components (Table 6.3).

Expertise required: The simple, user-friendly interface allows this tool to be used by non-experts who have a good understanding of the building envelope.

Target user: Engineers, designers, researchers.

Reference: Grunewald, Kolarik, Nemcova, & Weiss (2018).

Table 6.3 DELPHIN simulation tool.

	DELPHIN
Dimension	1D–2D
HAM model: moisture driving potential	Vapour pressure for vapour diffusion, moisture content or water pressure for liquid diffusion (user choice)
Moisture transfer	Vapour diffusion, liquid transport, moist air movement within the envelope
Numerical method	Finite Control Volume
Boundary conditions at the surface	Combined constant coefficient for long-wave radiation and convection
Wind-driven rain	Yes
Materials properties	Function of relative humidity, thermal conductivity and vapour diffusivity function of temperature

6.2.2.3 *BSim*

BSim is an integrated design tool developed by the Danish Building Research Institute (SBi) at Aalborg University. The software allows a building data model to be created and shared using the multiple tools available in the BSim family. These include whole building indoor climate, energy and daylight assessments, as well as prediction of the 1D envelope's hygrothermal behaviour. The moisture transfer mechanisms considered are vapour diffusion and convection; liquid transfer within the building component is not included.

The ability to import the 3D geometry from CAD software makes BSim highly suitable for a multitude of uses. It also offers a catalogue of building components and internal schedules that users can simply drag and drop to apply to the model, which makes BSim easy to use by non-experts. In regard to the accuracy of the hygrothermal calculation, BSim defines the materials properties as constant, thus overlooking their dependency on temperature and moisture variations, except for vapour permeability, thus generating results that are less accurate and realistic than those from other competing products (Table 6.4).

Expertise required: Users must have some general knowledge of building design and how buildings behave thermally in order to create the building model.

Target user: Engineers, researchers and students.

Reference: Rode & Grau (2003).

6.2.2.4 *IDA-ICE*

IDA-ICE (Indoor Climate and Energy) is an integrated building simulation tool that combines simulation of building envelope, systems and controls. The scope of this software is broad and it can compute several phenomena, such as integrated airflow, thermal models, CO_2, moisture transfer, vertical temperature gradients, daylight and

Table 6.4 BSim simulation tool.

	BSim
Dimension	1D–whole building
HAM model: moisture driving potential	Vapour pressure
Moisture transfer	Convection, vapour diffusion
Numerical method	Finite Control Volume
Boundary conditions at the surface	Combined constant coefficient for long-wave radiation and convection – it is possible to separate the calculations
Wind-driven rain	No
Materials properties	Constant (except water vapour permeability, which depends on relative humidity)

energy. The moisture transfer accounts for vapour diffusion and moist air movement within the envelope; liquid transport is not included. IDA-ICE is able to compute wind- and buoyancy-driven air infiltration through leaks and openings using an integrated airflow network model.

The broad scope of this software reduces the potential application and accuracy of hygrothermal assessments. It does not account for wind-driven rain, water deposition on the facades and variability of the materials properties as a function of temperature and relative humidity. The main advantage of this software is the two-layer approach to the simulation. The graphical interface is intuitive and easy for non-expert users to navigate, and it presents both 3D representations and tabular values. However, IDA-ICE's modular structure means that the model equations can be adapted and the tool's capabilities can be expanded by adding new extensions, thus accommodating the needs of more experienced and expert users (Table 6.5).

Expertise required: HVAC engineers and designers without specific expertise in building simulation, as well as advanced users who can build their own systems and equations.

Target user: HVAC designers, sustainable design engineers, experts in building simulations, researchers.

Reference: Kurnitski & Vuolle (2000); Sahlin et al. (2004); (Kalamees, 2004).

6.2.2.5 hygIRC – 1D-2D

hygIRC is a commercial software developed by the National Research Council Canada as an enhanced version of the LATENITE model. This tool accounts for HAM (liquid and vapour) transport, variability of the material properties with relative humidity, airflow through the building component, water phase-change effects, radiation coefficient, and moisture infiltration within the envelope. The latter can be expressed as a function of time at any location on the component, based on empirical results obtained in laboratory tests.

Table 6.5 IDA-ICE simulation tool.

	IDA-ICE
Dimension	1D
HAM model: moisture driving potential	Vapour density
Moisture transfer	Vapour diffusion, moist air movement within the envelope
Numerical method	Finite Control Volume
Boundary conditions at the surface	Separated convective coefficient (dependent on temperature and slope) and long-wave radiation (view factors)
Wind-driven rain	No
Materials properties	Constant – except water vapour permeability function of relative humidity

Table 6.6 hygIRC-1D/2D simulation tool.

	hygIRC
Dimension	1D–2D
HAM model: moisture driving potential	Moisture content and vapour pressure
Moisture transfer	Vapour diffusion, moist air movement within the envelope, liquid transport
Numerical method	Finite Control Volume
Boundary conditions at the surface	Constant coefficient for radiation and convection
Wind-driven rain	Yes
Materials properties	Function of relative humidity

A particular interesting feature of hygIRC is its ability to compute the indoor climate from the whole building HAM model, thus capturing the interactions between the envelope and the adjacent room. The major drawback is the exclusion of the influence of temperature on the building material properties, which can have a significant impact on the results. In addition, the user is required to model the envelope geometry through a pre-processor that requires the wall to be divided into a number of vertical and horizontal layers, a task that relies on the user's personal expertise and knowledge of the physical problem (Table 6.6).

Expertise required: General knowledge of building design and detailed knowledge of envelope hygrothermal performance (2D version).

Target user: Designers, engineers.

Reference: Defo (2018).

Table 6.7 trnsys ITT simulation tool.

	trnsys ITT
Dimension	1D–whole building
HAM model: moisture driving potential	Vapour pressure
Moisture transfer	Vapour diffusion, liquid transport
Numerical method	Finite Elements Model
Boundary conditions at the surface	Convective coefficient can be constant or temperature-dependent, long-wave radiation through view-factors
Wind-driven rain	No
Materials properties	Function of relative humidity and temperature

6.2.2.6 trnsys ITT

TRNSYS is a commercial software developed for thermal whole building simulation purposes which has been constantly improved and updated to integrate advanced knowledge in building physics. It has a modular structure, whereby users can choose different components, called types, and connect them to form a logical flow of information to create the simulation; TRNSYS ITT is one of these types. The general software allows the user to decide whether to employ the simplified model based on the moisture penetration depth, or to connect the TRNSYS ITT type. This type includes a more refined HAM model, accounting for moisture diffusion, liquid and vapour transport and phase changing impacts, all of which are combined with a realistic and accurate whole building model. However, the graphical interface is not easy to learn without specific training, it relies on the expertise of the users to link one type to the other correctly, and requires deep knowledge of building physics to interpret the mathematical model of each type (Table 6.7).

Expertise required: Experience with building simulation language, in-depth knowledge of building physics.

Target user: Researchers, building simulation experts.

Reference: Perschk (2000).

6.2.2.7 1D HAM

This software is a one-dimensional HAM calculation tool developed by the Lund–Gothenburg Group for Computational Building Physics to assess the hygrothermal behaviour of a multilayer porous building envelope exposed to the outdoors. The model allows heat to be transferred through conduction, convection and latent heat effects; moisture transport occurs by vapour diffusion and convection;

Table 6.8 1D-HAM simulation tool.

	1D-HAM
Dimension	1D
HAM model: moisture driving potential	Humidity by volume (function of relative humidity and humidity at saturation)
Moisture transfer	Convection, vapour diffusion
Boundary conditions at the surface	Modelled as a fictitious thin layer, transparent for solar radiation and with no additional air flow resistance. Input needed: solar absorption factor, thermal resistance and vapour diffusion resistance on both sides of the layer.
Materials properties	Constant

the liquid water transport mode is not included. Surface solar absorption is modelled through the use of a thin, fully transparent layer without air flow resistance adjacent to the external envelope material; the user needs to specify the surface thermal and vapour resistance of the two surfaces of this layer.

The major advantage of this software is its ability to capture and include air infiltrations through the envelope. The 1D-HAM model includes an airflow rate through the envelope, which is dependent on the pressure difference over the building envelope. The airflow rate is thus variable in time, and allows negative values to model a direction change of the flow. The major drawback of this software is that it requires a high level of experience and expertise to perform the assessments. Input data and assumptions are provided to the software through an external text file (.CLI extension), which is less intuitive than a fully integrated graphical interface (Table 6.8).

Expertise required: Deep understanding of building physics and building simulation processes.

Target user: Users with experience in building simulations, researchers.

Reference: Kalagasidis et al. (2007).

6.2.3 Research software and freeware: simulink environment

Several freeware and research software programs use the graphical programming languages Simulink and Matlab. These software are divided into blocks, such as the building component and the adjacent thermal zone, which have predefined libraries of calculations that compute the HAM transfer within a specific subsystem. These sub-systems are grouped into different categories related to the building envelope, the thermal zone, the systems (HVAC), the boundary conditions (weather file) and gains. These tools are open source software, allowing for implementation and integration by users. Despite the high level of customisation and flexibility that this offers, the complexity of the platform on which it

Table 6.9 Simulink simulation tools.

	HAM-Tools	**HAMFitPlus**	**HAMlab (HAMBASE)**
Dimension	1D	1D – 2D – whole building	1D – 2D – whole building
HAM model: moisture driving potential	Vapour pressure for vapour, suction pressure for liquid and moist air pressure for airflow	Relative humidity	Vapour pressure
Moisture transfer	Vapour diffusion, liquid transport, moist air movement within the envelope	Vapour diffusion, liquid transport, moist air movement within the envelope	Vapour diffusion, moist air movement within the envelope
Numerical method	Finite Control Volume	Finite Element Method	Finite Control Volume
Boundary conditions at the surface	Combined constant coefficient for long-wave radiation and convection – possible to perform separate calculations with constant or temperature-dependent convection and view factors	Combined constant coefficient for long-wave radiation and convection	Separated constant convective coefficient and long-wave radiation (sphere approximation)
Wind-driven rain	Simplified method	Yes	Yes
Materials properties	Function of temperature and moisture	Function of relative humidity dependent, vapour permeability function of temperature	Constant – but possible integration of COMSOL to express dependence on relative humidity and temperature

is based makes these tools more suitable for research purposes than for commercial use (Table 6.9).

Expertise required: Deep understanding of the HAM processes and implications for the building, as well as knowledge of the Simulink language.

Table 6.10 Other simulation tools.

	UMIDUS	Clim 2000	XAM	NPI
Dimension	1D	1D	1D	1D
HAM model: moisture driving potential	Vapour pressure or temperature and moisture volumetric content	Vapour pressure	Moisture content	Relative humidity
Moisture transfer	Vapour diffusion, liquid transport	Vapour diffusion	Vapour diffusion	Vapour diffusion, liquid transport
Numerical method	Finite Control Volume	Finite Control Volume	Finite Control Volume	Transfer Function
Boundary conditions at the surface	Convective coefficient can be constant or temperature-dependant, long-wave radiation through view-factors	Combined constant coefficient for radiation and convection, or separated view factors for radiation, temperature-dependent convective coefficient	Combined constant coefficient for long-wave radiation and convection	Combined constant coefficient for long-wave radiation and convection
Wind-driven rain	No	No	No	No
Materials properties	Function of relative humidity; also possible to choose function of temperature	Constant	Constant	Constant, except for vapour permeability and moisture diffusivity (function of temperature and relative humidity)

Target user: Researchers, experts in building simulations.
Reference: de Wit (2006); Kalagasidis et al. (2007); Kalagasidis, Rode, & Woloszyn (2008); Schijndel (2007); Tariku, Kumaran, & Fazio (2010).

6.2.4 *Other software*

Reference: Iwamae, Hanibuchi, & Chikada (1999); Koronthályova et al. (2004); Koronthályová (1998); Mendes & Philippi (2005); Mendes, Oliveira, & Santos (2003); Woloszyn, Rusaouen, & Covalet (2004) (Table 6.10).

6.3 Concluding remarks

Transient hygrothermal assessments require the use of specific software that can solve the differential equation of the heat, moisture and air transport through the building envelope and exchange with the internal and external environments. Over the years, numerous stand-alone or integrated pieces of software have been developed to aid hygrothermal computation, and a number of options are currently available. Each relies on a different hygrothermal model, with different levels of accuracy, assumptions, strengths and criticalities. Unfortunately, there is no single 'best' tool. The choice of which one to use must be carefully analysed based on a multicriteria approach. The main points to consider are:

- Purpose of the analysis: The tool must be able to provide sufficient information to support decision-making at the design stage. A software tool that can only provide general feedback at the concept design stage will probably not be suitable for assessing the specific hygrothermal performance of a material.
- Level of expertise needed: Different tools require different levels of experience in building simulation and/or knowledge of building physics; obviously, tools that require high levels of expertise will be more suitable for advanced studies.
- Accuracy of the assumptions: Depending on the purpose, a different level of accuracy may be considered acceptable.
- Flexibility: Most commercial software is not cheap, so it is important to consider the flexibility of the tool and its ability to be adapted to different purposes. One important aspect, for example, may be the capacity for integration into a BIM environment or to perform both 1D and whole building simulations.

References

ASTM. (2016). *E3054/E3054M: Standard Guide for Characterization and Use of Hygrothermal Models for Moisture Control Design in Building Envelopes.*

Blocken, B. (2004). *Wind-driven rain on buildings. Measurements, numerical modeling and applications. Wind-driven rain on buildings. Measurements, numerical modeling and applications.* Catholic University of Leuven.

Burch, D.M., & Chi, J. (1997). *MOIST: A PC program for predicting heat and moisture transfer in buildingenvelopes.*

Chung, D., Wen, J., & Lo, L. J. (2019). Development and verification of the open source platform, HAM-Tools, for hygrothermal performance simulation of buildings using a stochastic approach. *Building Simulation.*

de Wit. (2006). *Hambase: heat, air and moisture model for building and systems evaluation.*

Defo, M. (2018). *hygIRC—Review of the Implementation of a Hygrothermal Simulation Model.*

Delgado, J., Ramos, N. M., Barreira, E., & Freitas, V. P. D. (2010). A critical review of hygrothermal models used in porous building materials. *Journal of Porous Media, 13*(3).

Delgado, J.M., Azevedo, A.C., & Guimarães, A.S. (2019). *Interface Influence on Moisture Transport in Building Components: The Wetting Process.*

Delgado, J.M., Barreira, E., Ramos, N.M., & Freitas, V.P.D. (2012). *Hygrothermal numerical simulation tools applied to building physics.*

Grunewald, J., Kolarik, J., Nemcova, V., & Weiss, D. (2018). Using the PASSYS cell for model-to-model comparison of hygrothermal building envelope simulation tools. In *7th International Building Physics Conference.*

Hagentoft, C.-E. (2001). *Introduction to building physics.*

Hens, H.S. (1996). *Heat, Air and Moisture Transfer in Insulated Envelope Parts: Task 1, Modelling, Final Report, Laboratorium Bouwfysica.*

Hens, H. S. (2017). *Building physics-heat, air and moisture: fundamentals and engineering methods with examples and exercises.* John Wiley & Sons.

ISO EN 13788. (2012). *Hygrothermal Performance of Building Components and Building Elements—Internal Surface Temperature to Avoid Critical Surface Humidity and Interstitial Condensation—Calculation Methods* (Vol. 13788). Geneva: International Organization for Standardization.

ISO EN 15026. (2007). *Hygrothermal performance of building components and building elements—assessment of moisture transfer by numerical simulation.*

Iwamae, A., Hanibuchi, H., & Chikada, T. (1999). A Windows-based PC-software to design thermal environment in residential houses. In *Proceedings of Building Simulation.*

Kalagasidis, A.S., Rode, C., & Woloszyn, M. (2008). HAM-Tools—a whole building simulation tool in Annex 41. *IEA ECBCS Annex, 41.*

Kalagasidis, A. S., Weitzmann, P., Nielsen, T. R., Peuhkuri, R., Hagentoft, C. E., & Rode, C. (2007). The international building physics toolbox in simulink. *Energy and Buildings, 39*(6), 665—674. Available from https://doi.org/10.1016/j.enbuild.2006.10.007.

Kalamees, T. (2004). IDA ICE: the simulation tool for making the whole building energy and HAM analysis. *Annex, 41*, 12—14.

Koronthályová, O. (1998). Non-steady model for calculation of indoor air relative humidity. *Building Research Journal, 46*, 201—211.

Koronthályova, O., Mihálka, P., Matiašovský, P., Križma, M., Nürnbergerová, T., Jerga, J., ... Darula, S. (2004). Model for complex simulation of ham-transfer in a single thermal zone building. *Building Research Journal, 52*(4), 199—217.

Künzel, H.M. (1995). *One-and two-dimensional calculation using simple parameters. IRB-Verlag Stuttgart 65.*

Kurnitski, J., & Vuolle, M. (2000). Simultaneous calculation of heat, moisture, and air transport in a modular simulation environment. In *Proceedings of the Estonian academy of sciences engineering.*

Luikov, A. V. (1964). Heat and mass transfer in capillary-porous bodies. *Advances in Heat Transfer, 1*(C), 123—184. Available from https://doi.org/10.1016/S0065-2717(08)70098-4.

Mendes, N., & Philippi, P. C. (2005). A method for predicting heat and moisture transfer through multilayered walls based on temperature and moisture content gradients. *International Journal of Heat and Mass Transfer*, *48*(1), 37–51. Available from https://doi.org/10.1016/j.ijheatmasstransfer.2004.08.011.

Mendes, N., Oliveira, R., & Santos, G.D. (2003). Domus 2.0: a whole-building hygrothermal simulation program. In *Proceedings of Building Simulation* (pp. 863–870).

Nore, K., Blocken, B., Jelle, B. P., Thue, J. V., & Carmeliet, J. (2007). A dataset of wind-driven rain measurements on a low-rise test building in Norway. *Building and Environment*, *42*(5), 2150–2165. Available from https://doi.org/10.1016/j.buildenv.2006.04.003.

Pallin, S., Boudreaux, P., Shrestha, S., New, J., & Adams, M. (2017). *State-of-the-Art for Hygrothermal Simulation Tools*.

Perschk, A. (2000). *Gebäude-Anlagen-Simulation unter Berücksichtigung der hygrischen Prozesse in den Gebäudewänden*.

Philip, J. R., & De Vries, D. A. (1957). Moisture movement in porous materials under temperature gradients. *Transactions, American Geophysical Union*, 222. Available from https://doi.org/10.1029/tr038i002p00222.

Prowler, D., & Trechsel, H. (2008). Mold and moisture dynamics. In *Whole Building Design Guide 9*.

Rode, C., & Grau, K. (2003). Whole building hygrothermal simulation model. In *ASHRAE Transactions* (Vol. 109, pp. 572–582).

Sahlin, P., Eriksson, L., Grozman, P., Johnsson, H., Shapovalov, A., & Vuolle, M. (2004). Whole-building simulation with symbolic DAE equations and general purpose solvers. *Building and Environment*, *39*(8), 949–958, Elsevier BV. Available from https://doi.org/10.1016/j.buildenv.2004.01.019.

Salonvaara, M., Zhang, J., & Karagiozis, K. (2011). Enivornmental Weather Loads for Hygrothermal Analysis and Design of Buildings. In *ASHRAE RP-1325. Simulation studies and dataanalysis*.

Schijndel. (2007). *Integrated heat air and moisture modeling and simulation*.

Standard 160-2016. *Criteria for Moisture-Control Design Analysis in Buildings*. (2016). ASHRAE.

Straube, J., & Burnett, E. (2001). *Overview of hygrothermal (HAM) analysis methods. Moisture analysis and condensation control in building envelopes*. ASTM International.

Tariku, F., Kumaran, K., & Fazio, P. (2010). Transient model for coupled heat, air and moisture transfer through multilayered porous media. *International Journal of Heat and Mass Transfer*, *53*(15–16), 3035–3044. Available from https://doi.org/10.1016/j.ijheatmasstransfer.2010.03.024.

Woloszyn, M., & Rode, C. (2008). Tools for performance simulation of heat, air and moisture conditions of whole buildings. *Building Simulation*, 5–24. Available from https://doi.org/10.1007/s12273-008-8106-z.

Woloszyn, M., Rusaouen, G., & Covalet, D. (2004). Whole building simulation tools: Clim2000. *IEA Annex*.

Building codes and standards 7

Worldwide, countries are increasingly aiming to achieve complete decarbonisation of the built environment by 2030, resulting in significant policy changes. From a building design perspective, one of the major implications is the minimum energy performance requirement for buildings, which has led to rapid diffusion of so-called high-performing building envelopes. At the same time, increasingly high energy efficiency requirements are revolutionising the hygrothermal (heat and moisture) performance of buildings. With the exception of a few instances in which these requirements are met through natural ventilation and a permeable envelope, they more commonly rely on increased airtightness levels, sealed indoor environments and highly insulated envelopes. Yet these measures are not usually accompanied by adequate ventilation strategies or construction practices, which leads, respectively, to under-ventilated indoor areas that are unable to dissipate the additional moisture, and a high rate of occurrence of thermal bridges, which favour condensation and mould growth in the build-ups.

Although much attention has been paid to the thermal performance of the envelope, its response to and interaction with moisture have been overlooked. Currently, various countries are addressing these risks and implementing specific indications, provisions and assessment tools to help designers to provide optimal envelope solutions in regard to hygrothermal performance.

This chapter explains how relevant hygrothermal standards adopt different hygrothermal models to assess moisture and condensation risk and provides a snapshot of the current regulatory state of the art in regard to moisture, condensation and associated risks for public health in the built environment. The focus is on Europe, America and Australia. This enables a comparative analysis of the building codes adopted in Europe and America, which are leaders in research and standardisation through the ISO and ASHRAE standards, and the special case of Australia, where the National Construction Code (NCC) included moisture management provisions to address the condensation issue for the first time in 2019. Australia represents an interesting case study due to the size of its territory and the diversity of its climates. Here, moisture management strategies and building code implementation build on research and knowledge developed in Northern Europe and America, where the focus is on cold and temperate climates. An analysis of how the Australian policy has adapted this knowledge for warm climates reveals some gaps in the current regulations and standards, as well as opportunities for further research.

Moisture and Buildings. DOI: https://doi.org/10.1016/B978-0-12-821097-0.00008-4

The following codes are considered in the comparative analysis:

1. European/English code (BS/ISO)
2. American National Code (ANSI/ASHRAE)
3. Australian National Code of Construction (NCC)—Building Code of Australia (BCA).

7.1 European/British building regulations

The European ISO standards comprise a set of norms, guidelines and indications designed by all relevant stakeholders through a transparent, open and consensual process. The standards can be adopted as regulatory mechanisms at the regional level, with appropriate adaptations to reflect different climates and social–cultural contexts. This chapter uses the British regulatory system as an example to contextualise the international standard. Although British building regulations have long experience in dealing with moisture and condensation assessment, condensation-related issues continue to be under discussion.

The British Building Code states that 15% of British houses are still affected by mould growth and condensation to some extent (Building Regulations and Approved Documents, n.d.). However, this figure may be much higher according to a recent World Health Organisation report, which suggests that up to 45% of European houses might suffer from mould and moisture-related damage (Clark et al., 2004). Whatever the correct figure, it is clear that moisture is still an issue for buildings. This might be attributable to inadequate understanding of condensation in the building sector, which leads to underestimation of its effects. Successful moisture management not only involves the adoption of appropriate design strategies but also requires attention to the building's construction and operation, which are extremely difficult to control and regulate. However, the British building regulations remain one of the most complete in Europe. The following analysis discusses the main documents and standards that relate to moisture issues and design, identifying those shortcomings that need to be addressed to mitigate the condensation-related risks.

The building regulations set minimum standards for design, construction and alterations for virtually all buildings. The approved documents, developed by the UK government and approved by the Parliament, also contain guidance on how to meet the building regulations. The so-called Part C on 'Site preparation and resistance to contaminants and moisture' (Part C, DCLG, 2013) of the British Building Regulations is relevant to condensation, and it refers to the BS 5250:2011, as well as the BS EN ISO 13788:2012 and BS EN 15026:2007. It applies to England, Scotland and Wales, and defines the requirements for preparation of the site, resistance to contaminants (C1) and resistance to moisture (C2). Sections 4, 5 and 6 tackle the problem of moisture using a holistic approach, acknowledging that mould and condensation may be due to different causes, such as:

- moisture emanating from the ground or groundwater;
- precipitation and wind-driven spray;
- spillage of water from or associated with sanitary fittings and fixed appliances;
- surface or interstitial condensation.

Approved document F (DCLG, 2010) also includes design guidance for the provision of minimum ventilation rates and for reducing the risk of surface mould growth. However, these guidelines are very general and are aimed primarily at new buildings rather than retrofit, which presents a more complex problem. Further, the criteria for mould growth consider only conditions that are typical for internal surfaces (May & Sanders, 2016).

7.1.1 BS 5250:2011 - Code of practice for control of condensation in buildings

This standard (BS 5250, 2011) provides a high-level description of the physical phenomena involved in hygrothermal processes within the built environment. It describes condensation causes and effects, especially in regard to the environmental health risk of mould, early biodeterioration and durability of materials, and possible structural failures. It also provides a set of regulatory and informative guidelines for the design stage that can help to reduce the risk of condensation. These can be grouped into the following six categories:

1. Design to avoid moisture: overview of the problems related to moisture and condensation in buildings, and information for designers on how to reduce or avoid the risks.
2. Diagnosis of dampness problems: cause of dampness, recognition of hygroscopic effects and rising dampness.
3. Guidelines for condensation assessment with reference to other codes, including BS EN ISO 13788:2012 *Hygrothermal Performance of Building Components and Building Elements. Internal Surface Temperature to Avoid Critical Surface Humidity and Interstitial Condensation* (BS EN ISO 13788, 2012), BS EN 15026:2007 *Hygrothermal Performance of Building Components and Building Elements. Assessment of Moisture Transfer by Numerical Simulation* (BS EN 15026, 2007).
4. Properties of construction materials: typical conductivity and vapour resistivity values.
5. Application of design principles on floors, walls, roofs, and junctions, with examples of build-up and joint resolution to avoid or reduce thermal bridges and risks of condensation.
6. Guidelines for builders and owners.

BS 5250:2011 applies to all building types, and it covers both the design of new buildings and renovation of existing ones. It provides recommendations and guidelines for the correct management of hygrothermal issues during the design stage, especially in regard to condensation. The goal of the design guidelines is to avoid high moisture levels in the building envelope, which can easily lead to condensation-related issues, such as mould growth or material deterioration.

The code acknowledges the increasing relevance of condensation issues in the built environment, which it sees as a possible consequence of the increasing requirements for energy efficiency that encourage hyper-insulated and more airtight envelopes. The stricter thermal requirements are not usually accompanied by updated ventilation strategies, which magnifies the potential for condensation if the envelope is not designed correctly. A major limitation of this code is its limited focus on the winter scenario, when buildings are heated and the temperature is higher indoors than outdoors. Although this represents the traditional British climate, the code itself acknowledges the impact of global warming and the potential modification of the standard scenario of use. Annex A elaborates on the 'unforeseen risk of interstitial condensation' caused by the increasing use of air-conditioning (AC) during summer months, which reverses the hygrothermal flux directions from outward to inward. In their present form, the British Building Regulations cannot fully capture this situation, resulting in envelopes that are ill-prepared to manage the hygrothermal pressure.

7.1.1.1 Content and scope

This code provides recommendations and guidelines to avoid issues related to high moisture levels and condensation, based on construction typologies commonly used in the United Kingdom. It offers a dual approach to hygrothermal risk assessment and design verification: it provides both calculation methods and specific references to other codes (specifically BS EN ISO 13788:2012) where the methods are described in detail, as well as high-level design criteria for floors, wall and roof build-ups, including prescriptive examples of reference details (Annexes F, G and H). The high-level provisions presented in the body of the code can be summarised as follows:

- Material with the highest and lowest vapour resistance should be located on the warm and cold sides, respectively.
- Minimise gaps between insulation panels and thermal bridges in general to avoid localised low surface temperatures with a consequent high risk of condensation.
- Avoid warm air in cold cavities.
- Air and vapour control layers to be lapped at least 50 mm and sealed.
- A void should be formed behind the internal surface finish to enable services to be installed without compromising the air and vapour control layer.
- In the case of remediation, increased heating or the addition of insulation is suggested.
- Keep materials dry before and during construction; water trapped inside construction materials can take a long time to dry out.

The code includes a series of annexes that further clarify the recommendations. For example, Annex A, B and C deal with the general theoretical framework of moisture, dampness and moisture content of air to provide the background knowledge necessary to understand the rationale behind the specific guidelines. Annex D addresses surface condensation assessment. It describes the different methods available to assess the risk of condensation and provides recommendations for their application, distinguishing between:

1. D.3.1.3—Steady state BS EN ISO 13788:2012—1 Dimensional analysis—simplified approach
2. D.3.1.4—Transient state BS EN 15026:2007—1 Dimensional analysis
3. D.3.1.5—Computational fluid dynamics—for more complex 2/3D models of heat, air and moisture movement.

The first method, described in detail in code BS EN ISO 13788:2012, is one of the simplest condensation assessment methods that can be applied during the design stage. Thus specific guidelines are provided to ensure correct application in relation to:

- conductivity and vapour resistivity of common construction materials;
- calculation/correction of monthly mean temperature and relative humidity (RH) for interstitial condensation calculations;
- moisture production rates in housing (based on number of occupants or activity);
- internal humidity classes subdivision depending on RH at internal temperature.

Annex D clearly acknowledges the limitations of this assessment: 'Designers should be aware that BS EN ISO 13788:2012 considers only the risks arising from the diffusion of water vapour through the building fabric; it does not take account of the much greater risk of condensation occurring as a result of air leakage, which transports water vapour through gaps, joints and cracks in the building fabric' (BS 5250, 2011). Annex E explains the relevant material properties and provides tables with numerical values of the most common building materials. Details of the prescriptive build-ups and components are included in Annexes F, G, H and J, which illustrate the application of the design principles previously explained. The code highly recommends following the prescriptive path, concluding that it is 'sufficient to provide robust designs to minimise moisture problems' (BS 5250, 2011). It is also stated that these prescriptive examples can be adopted without the need for any further analysis or verification. Annexes K and L deal with indoor spaces and the basic rules for correctly ventilating and heating the space without increasing the risk of mould occurrence. The code also acknowledges the importance of building construction and operation and provides additional guidelines for builders and owners/occupiers (Annexes M and N).

7.1.2 BS EN ISO 13788:2012—hygrothermal performance of building components and building elements. Internal surface temperature to avoid critical surface humidity and interstitial condensation

This standard (BS EN ISO 13788, 2012) is the main reference for hygrothermal assessment methods and their application. It contains the recommended risk assessment procedures for both mould growth and condensation, the indications for input selection and interpretation of the results.

7.1.2.1 Boundary conditions

Hygrothermal assessment methods require the definition of certain boundary conditions, namely, all those parameters that characterise the context of the analysis, such as internal and external climatic conditions or material properties. The code specifies how to set the boundary conditions, explaining that all the inputs are relevant and applicable to all the described methods.

These boundaries conditions include

- Material conductivity and water vapour resistance. The code refers to other international standards for these values as each country might have a national-specific table (e.g. BS 5250:2011 Annex E).

- External boundary conditions, such as monthly temperature and RH. These values are site-specific and should be taken from either other national standards or weather stations. Depending on the type of assessment, this might involve consideration of mean values or the average, taken over several years, of the lowest daily mean temperature in each year.
- Internal boundary conditions, such as temperature and RH. The temperature can be provided by thermal building simulation or, alternatively, calculated using the diagrams provided in Annex A, based on the external temperature and expected occupancy pattern.
- Surface resistance values can be obtained from the table depending on the direction of heat flow.

7.1.2.2 Assessment typologies, methods and results

The code describes both mould growth and condensation risk assessment. The condensation risk assessment method includes both surface and interstitial evaluation through the so-called *Glaser method*, which is a simplified method of calculating the amount of condensate deposit and/or evaporate over a 12-month period. It allows the quantification of a critical RH value to avoid condensation and condensation-related risks, while enabling a quick performance comparison of different design solutions and building components (refer to Chapter 3: Durability, condensation assessment and prevention). The mould growth risk evaluation method involves a two-step approach: first, establish the risk of condensation on the surface of a building component; and second, evaluate the risk of mould growth or biodeterioration of the material.

7.1.2.2.1 Surface condensation

The goal of this assessment is to define the critical value of the internal surface temperature ($\theta_{si,min}$) that can cause condensation and, consequently, determine the minimum thermal resistance of the building component (e.g. wall, floor or roof) that allows the internal surface temperature to be maintained above the critical threshold and, thus minimise the risk of condensation. Ideally, the identified threshold is also suitable for a more sophisticated 2D thermal analysis, allowing assessment of the risk of condensation in the presence of thermal bridges.

The assessment process aims to calculate 12 different values of minimum acceptable indoor surface temperature, one for each month. The critical month and surface temperature are determined through the calculation and comparison of the temperature factor at the internal surface (f_{Rsi}).

The 'minimum acceptable surface temperature' associated with the 'highest value of the temperature factor at the internal surface' is used to define the minimum envelope thermal resistance.

The code provides several variants of the same formula to enable the assessments of:

- Critical surface temperature to avoid mould growth. A critical surface RH of 0.8 is considered as input to avoid the risk of mould (unless more specific information is available). This calculation considers the monthly mean external temperature.

- Critical surface temperature to avoid the risk of corrosion. A critical surface RH of 0.6 is considered as input to avoid the corrosion. This calculation considers the monthly mean external temperature.
- Critical surface temperature to avoid the risk of condensation on low thermal inertia elements, including glass and frames. A critical surface RH of 1 is considered to avoid corrosion in metal frames or rot in wooden ones. For this calculation, the external temperature must be defined as the average, taken over several years, of the lowest daily mean temperature in each year.

Refer to Chapter 3, Durability, condensation assessment and prevention, for a detailed description of this method.

7.1.2.2.2 Interstitial condensation—Glaser method

This method is used to quantify the maximum amount of accumulated moisture due to interstitial condensation caused by vapour flow through the building element. Due to the nature of the calculation, this has to be considered as an assessment tool rather than an accurate prediction tool (refer to Chapter 3, Durability, condensation assessment and prevention, for a detailed description of the Glaser method).

7.1.2.2.3 Drying potential of building components

The code identifies the Glaser method as a suitable tool to assess the drying potential of building components, which can be important when external walls and roofs are wrapped with high vapour resistant layers (water vapour permeability of material with respect to partial vapour pressure $s_d > 2$ m) such as waterproofing membranes, vapour barriers or coatings. This method assumes that building materials can be wet due to built-in moisture, rain during construction, leaks from services, defects in a waterproof layer or previous interstitial condensation problems. This extra moisture content is accounted through an additional 1 kg/m^2 of water concentrated at the centre of the layer. The Glaser method is used to evaluate the evaporation rate for each month to quantify the time necessary to remove all the additional moisture from the building component.

This assessment can generate three different outcomes:

- Drying out within 10 years without condensation in other layers. Depending on the time it takes for the water to evaporate, assess the risk of degradation to the layer containing the excess moisture.
- Drying out within 10 years with temporary condensation in other layers. Depending on the time it takes for the water to evaporate, assess the risk of degradation to the layer containing the excess moisture and to all other layers that show condensation issues.
- Drying time exceeds 10 years. The code does not specify any action.

7.1.2.3 Limitations

The code clearly acknowledges the assumptions and limitations of these suggested calculation methods, which can be summarised as follows:

- The methods assess the risk of interstitial condensation considering the hygrothermal flux as mono-directional. The one-dimensional calculation allows the hygrothermal behaviour of the building component to be modelled only as perpendicular to its surface; therefore it cannot be used to analyse complex build-ups or thermal bridges, which would require a two-dimensional approach.

- The indicated methods are based on a steady-state approach, which uses monthly means for temperature and RH. The real boundary conditions are not constant over a month, and the methods do not take into account the diurnal variations of the internal and external climate.
- The methods overlook the contribution of air movement through the envelope; thus the reliability and accuracy of the results are greater for airtight envelopes.
- The moisture-buffering properties of building materials are not considered, nor are the effects of additional moisture content within material (e.g. water absorbed during construction) or the susceptibility of material properties to moisture. Consequently, this approach provides a more robust analysis for lightweight structures, characterised by less permeable materials, with lower moisture-buffering capacity.
- Capillary suction and liquid moisture content, including penetration by rain and groundwater, are not considered, thus underestimating the source of potential moisture infiltration that may lead to higher risk of condensation.
- The standard neglects the dependency of water permeability of air to temperature and barometric pressure introducing an additional source of errors.
- The surface condensation assessment is defined for the internal surface only and is not considered reliable for evaluating the risk of condensation on external surfaces due to the overly simplified boundary conditions.

The indicated methods are applicable only with respect to the limitations above and, as stated in the introductory section of the code, the calculations have been proved to lead to design 'well on the safe side' (BS EN ISO 13788, 2012).

7.1.3 BS EN 15026:2007—Hygrothermal performance of building components and building elements. Assessment of moisture transfer by numerical simulation

This code (BS EN 15026, 2007) aims to standardise the use of hygrothermal software for condensation risk assessment and for the evaluation of the hygrothermal performance of building components. In particular, it focuses on software used to perform transient simulations and predict the one-dimensional heat and moisture transfer within a multilayer building envelope component subjected to non-steady climate conditions on either side. This code describes the software core equations, it identifies standardised input data and suitable display of results. Transient assessment methods are more accurate and specific compared to the method described by BS EN ISO 13788:2012. Indeed, this code addresses the issue of heat (both sensible and latent) and moisture storage phenomena (including absorbed water) and moisture transport by liquid transport (including surface diffusion and capillarity), allowing for a transient iterative calculation based on hourly data.

This method can yield different outputs, including:

- drying of initial moisture uptake of the construction phase;
- moisture accumulation by interstitial condensation due to diffusion in winter;
- summer condensation due to migration of moisture from outside to inside

- moisture penetration due to exposure to driving rain;
- exterior surface condensation due to cooling by longwave radiation exchange
- moisture-related heat loss by transmission and moisture evaporation.

7.1.3.1 Boundary conditions

Like BS EN ISO 13788:2012, this code specifies the boundary conditions that must be considered during the hygrothermal assessment in relation to material properties and internal and external climates. The internal climate must account for the worst-case scenario, which includes daily and seasonal temperature fluctuations and changes in building use. These modifications are introduced to account for possible calculation errors and uncertainties embedded in the modelling, as a real-case scenario might not perfectly align to the model's limitations. Whereas the determination of internal boundary conditions is similar to BS EN ISO 13788:2012, the use of external averaged conditions is not appropriate for this assessment method. Indeed, the external climate must be representative of the specific building's location and it must represent the extreme case. Considering that, in most moisture applications, a 'once-in 10-years' failure rate is usually considered acceptable, the most appropriate time series for the external weather data should include 10 years of measured data.

Alternatively, the code accepts the use of:

- a design reference year, which is a file created *ad hoc* to represent the most severe conditions likely to occur once every ten years, or
- a reference year created by applying a temperature shift to mean values.

Annex B of BS EN 15026:2007 gives a set of guidelines, also referencing EN ISO 15927-4:2005 (EN ISO 15927-4, 2005), on how to create these files and the minimum parameters that they should include, namely:

- vapour pressure or any other humidity parameter that can be used to calculate vapour pressure
- dry bulb temperature
- global and diffuse solar radiation
- sky temperature
- wind speed and direction
- total atmospheric pressure
- precipitation (rain, snow, drizzle)

A complete weather file must include the hourly value of the above-specified parameters, to allow the iterative, transient calculation. Compared with BS EN ISO 13788:2012, this analytical model takes into account more complex ways in which the material can transfer heat and vapour, such as radiative and convective heat transfer, capillarity and diffusion moisture transport.

7.1.3.2 Results

The code also provides guidelines for reporting the results. In particular, it prescribes the minimum content that must be included in the report and an indicative list of essential outputs that the software needs to be able to provide. The code requires the reporting of transient distribution, mean and peak values of the main hygrothermal parameters (temperature, heat flow, water content, moisture flow, RH and vapour pressure), and the boundary conditions used in the assessment. However, the code fails to provide specific information and guidelines on the interpretation of the numerical results. BS EN 15026:2007 does not give any specific information, nor does it reference other relevant codes or documents. It merely mentions a non-specific 'post-process model to assess mould or algae growth, rot, corrosion' without further elaboration. This lack of specificity raises the same issue as was discussed in relation to BS EN ISO 13788:2012; that is, the urgent need for a standardised matrix for the interpretation of results and the current need for a professional 'expert' to provide the relevant input.

7.1.3.3 Limitations

The code acknowledges the limitations of the transient assessment method, especially in regard to the simplification of the model necessary to run the simulation in the virtual environment. For this reason, this code should not be used in cases where:

- convection takes place through holes and cracks;
- two-dimensional effects play an important part (e.g. rising damp, conditions around thermal bridges, the effect of gravitational forces);
- hydraulic, osmotic and electrophoretic forces are present;
- daily mean temperatures in the component exceed 50°C.

The hygrothermal equations specified by the code are based on the conservation of energy and moisture and contain the following assumptions:

- constant geometry, no swelling and shrinkage;
- the absence of chemical reactions;
- latent heat of sorption is equal to latent heat of condensation/evaporation;
- no change in material properties by damage or ageing;
- local equilibrium between liquid and vapour without hysteresis;
- moisture storage function is not dependent on temperature;
- temperature and barometric pressure gradients do not affect vapour diffusion.

7.1.4 General comments

The European code provides three levels of guidance: prescriptive provisions, high-level design criteria indications and assessment methodology, including specific guidelines for designers, builders and occupants. However, although it tackles the issue from all three perspectives of design, construction and operation, it has some limitations that reduce its accuracy and potential usefulness.

The code specifically focuses on cold climates and, although it acknowledges the role of AC and warm climates in moisture-related problems, it does not address these problems or provide high-level indications on how to do so, failing to offer robust and coherent prevention strategies. In addition, the provisions for existing buildings and renovations are limited in comparison to those for new buildings and, even in the case of recommendations for new buildings, the types of build-ups and components proposed may be out of date or not relevant for new construction techniques. This narrow focus significantly limits the code's scope and applicability, thus undermining its intended purpose.

BS EN ISO 13788:2012 and BS 5250:2011 contain guidelines that enable the designer to undertake the required assessments of hygrothermal performance of building components. However, although the calculation method and input data are clearly described, the code lacks a clear matrix or detailed guidelines in regard to the interpretation of the numerical results, often referring to other inadequately specified 'regulatory requirements and other guidance in product standard'. The same observations can be made for BS EN 15026:2007, which lacks a clear set of targets and an evaluation matrix to interpret the outputs. Accordingly, designers or certifiers can only base their professional advice on previous experience. This may lead to misuse by professionals who lack the necessary experience, implicitly highlighting the need for 'expert' input.

From another perspective, a certain level of difficulty in the interpretation of the data is inevitable given the potential for unreliability of the results due to the complexity of the physical mechanisms that regulate moisture transfer and condensation in the built environment, the simplifications of the physical model and the limitations of the Glaser method. The long list of assumptions and limitations associated with the code raises questions about its reliability. The use of averaged climatic conditions reduces the accuracy of this approach under the actual conditions in which the building operates. Further, since the problems of condensation and mould are often related to thermal bridges, 1D nature of this method does not allow a complete risk assessment to be performed. Finally, an additional layer of complexity is added by the lack of a clear model that correlates the environmental parameters with the risk of mould growth and material degradation. This issue is exacerbated by the availability of a wide range of products on the market and their differential behaviour when exposed to specific temperatures and RH.

7.2 American building code

Most of the American states have adopted the International Building Code (IBC, 2018), in full or in a modified version, as a National Building Code. For this reason, the IBC 2018 can be said to represent the American legislative framework for analytical purposes. Like the British code document C and the BS 5250:2011, the IBC 2018 contains a number of prescriptive provisions. Chapter 12 of IBC 2018, Interior Environment, sets out minimum provisions for the interior of buildings—the occupied environment. This section directly regulates ventilation, lighting and space heating in conjunction with the International Mechanical Code and the International Energy Conservation Code.

In regard to the hygrothermal performance of building components, the code provides recommendations for the location of air-permeable and air-impermeable insulation, in cases of unvented insulated rafter assemblies, and of vapour retarders for roofs and external walls. Specific recommendations are given for different permeability classes of the vapour barrier in regard to the climatic zone. The code divides vapour retarders into three classes: Class I: 0.1 perm or less, Class II: 0.1 < perm =/< 1.0 perm, Class III: 1.0 < perm =/< 10 perm. The permeability must be measured according to ASTM E96 (2016).

As stated in the code:

- Class I and II vapour retarders shall not be provided on the interior side of frame walls in Zones 1 and 2. Class I vapour retarders shall not be provided on the interior side of frame walls in Zones 3 and 4 other than Marine 4. These provisions are specifically for hot and warm climates and aim to avoid condensation on the vapour retarder in cooling-mode, when the heating and vapour gradient is from outside to inside.
- Class I or II vapour retarders shall be provided on the interior side of frame walls in Zones 5, 6, 7, 8 and Marine 4. The exceptions are (1) basement walls; (2) below-grade portion of any wall; (3) construction where moisture or its freezing will not damage the materials; (4) conditions where Class III vapour retarders are required in the next point.
- Class III vapour retarders shall be permitted where any one of the conditions in table 1404.3.2 is met. Only Class III vapour retarders shall be used on the interior side of frame walls where foam plastic insulating sheathing with a perm rating of less than 1 is applied in accordance with table 1404.3.2 on the exterior side of the frame wall.

The code allows for modification if these prescriptive provisions cannot be met, namely, a performance design solution 'using accepted engineering practice for hygrothermal analysis'. However, the IBC 2018 does not specify or reference a tool or code that could be used for performance assessment in this context. In this regard, the ANSI/ASHRAE 160-2016 (ASHRAE 160, 2016) can be used to supply the additional information needed. This standard is gaining international acceptance as a benchmark and reference for hygrothermal performance assessment. This American National Standard is a national voluntary consensus standard developed under the auspices of the American Society of Heating, Refrigerating and Air Conditioning Engineers (ASHRAE), an international organisation devoted to indoor environments, with particular focus on topics related to heating, ventilation and cooling systems. Since IBC 2018 does not specifically reference the ASHRAE 160-2016 or any other previous versions, this standard remains voluntary, unless a legal jurisdiction has made compliance mandatory through legislation.

7.2.1 ASHRAE 160-2016—Criteria for Moisture-Control Design Analysis in Buildings

Standard ASHRAE 160-2016 specifies the performance-based design criteria for predicting and mitigating moisture damage on the building envelope, materials, components, systems and furnishings, depending on climate, construction type, and heating, ventilation and air-conditioning (HVAC) system operation. Like BS EN 15026:2007, it establishes standardised criteria to regulate the use of transient hygrothermal assessment software, based on an hourly calculation simulation method.

The use of analytical tools for assessment of condensation and mould growth risk or material deterioration is becoming increasingly common. However, the reliability of the assumptions, boundary conditions and conclusions needs to be established to guarantee robust results. Accordingly, this standard aims to provide a consistent framework with clear design assumptions and evaluation criteria that can be used to perform hygrothermal performance assessments. The ASHRAE 160-2016 describes a moisture assessment protocol that follows the performance-based design approach and can be applied to new buildings, retrofit and renovation of existing buildings. This protocol specifies the criteria for selecting analytic procedures, for defining inputs, and for evaluating and interpreting outputs.

7.2.1.1 Analytic procedures and inputs

The standard provides different methods of varying complexity for calculating the parameters necessary to set the boundary conditions and input values for the assessment. It includes charts and formulae as well as references to external codes. The use of more precise data from advanced simulation tools is encouraged to supply all the inputs needed for the transient assessment.

To provide a complete and accurate set of boundary conditions, the code provides guidance in regard to the following:

- initial design moisture content of building materials;
- internal loads—indoor temperature and indoor humidity;
- external loads—temperature, humidity, rain, air pressure and airflow.

7.2.1.1.1 Initial design moisture content of building materials

Building materials can uptake a significant amount of water during the construction phase. The ASHRAE 160-2016 accounts for this by assigning to each material an initial equilibrium moisture content, fixed for all materials at 80% of RH, with the exception of concrete, which is set at 90%. These are considered the highest moisture level that does not lead to mould growth. According to the standard, if insufficient care is taken to protect construction materials from wetting and limit the initial moisture conditions on-site, the design moisture content must be doubled (e.g. 2 × EMC90 for concrete and 2 × EMC80 for other materials).

7.2.1.1.2 Internal loads—indoor temperature and Indoor humidity

Transient hygrothermal assessments use hourly based indoor profiles for both temperature and humidity, which must be predicted or assumed during the design phase. In regard to temperature, the standard provides a simplified calculation method to generate the hourly profile, based on the outdoor temperature and HVAC system. This method is suggested when no design data are available from other energy simulation tools or other codes.

In regard to humidity, three different paths are indicated, depending on the available information and complexity of the process:

- Simplified method: here the design of indoor humidity is a function of the average daily outdoor temperature and level of occupancy. The method can be applied to residential and apartment buildings. This approach is similar to that indicated in European standard BS EN 15026:2007. However, it seems to generate inaccurate results for buildings with AC and dry climates (TenWolde, 2011a).

- Intermediate method: this methodology, based on the work of TenWolde and Walker (2001), generates the indoor design humidity from hourly based climatic parameters and the type of HVAC equipment, distinguishing between systems with and without dehumidification. When cooling and dehumidification equipment is not present or not operating, the method suggests using a mass balance equation. In this case, the calculation is done using the 24-hour running average outdoor vapour pressure to account for the moisture-buffering effect (TenWolde, 2011b). The code provides different variations of the formula to address the use of designed and non-designed ventilations, indoor design humidity with AC and dehumidification.
- Full parametric calculation: this method uses building modelling to account for the contributions of building thermal behaviour and ventilation.

7.2.1.1.3 External climatic loads

The standard provides clear guidelines on how to account for all the external climatic loads that can influence the hygrothermal performance of building components, such as wind, rain, temperature, humidity and solar radiation. Simulation software usually requires a so-called weather file, which collects all this information. The ASHRAE 160-2016 recommends the use of either a weather file with a minimum of 10 consecutive years of monitored weather data, or moisture design reference years weather data (MDRY), defined as the 10th-percentile warmest and 10th-percentile coldest years from a 30-year weather analysis.

The weather file must include hourly data on

- dry-bulb temperature;
- vapour pressure, dew point temperature, wet-bulb temperature, RH or humidity ratio;
- total solar insolation on a horizontal surface;
- average wind speed and direction;
- rainfall: in the absence of specific data from a full-scale test, the code assumes a default value of 1% of water penetration through the exterior surface; and
- cloud index.

As well as indicating the minimum content required for the weather file, the standard also includes simplified analytical procedures to calculate the rain load on walls. This calculation allows the designer to consider the impact of wind-driven rain that can deposit on the external layers of the wall and is based on the work of Lacy (1965). The underlying assumption is that the external cladding is not watertight, and a small amount of rain will deposit on the adjacent layer; in the absence of a tested value, the default penetration rate is set at 1% of the total external rain load.

7.2.1.2 Evaluation and use of outputs

The standard uses the hygrothermal performance criteria defined by the VTT growth model (refer to Chapter 4: Health and mould growth), which has been validated with experimental data for different materials with varying sensitivities to mould growth. It predicts mould growth risk as a function of time, surface temperature and surface RH. The standard sets the performance criteria based on critical mould growth thresholds, which have been found to be the strictest among the hygrothermal performance criteria (e.g. critical thresholds for durability).

The previous version of this standard, the ASHRAE 160-2009, included three additional pass-fail conditions to be met:

1. 30-day running average surface RH $< 80\%$ when the 30-day running average surface temperature is between 5°C (41°F) and 40°C (104°F);
2. 7-day running average surface RH $< 98\%$ when the 7-day running average surface temperature is between 5°C (41°F) and 40°C (104°F);
3. 24-day running average surface RH $< 100\%$ when the 24-h running average surface temperature is between 5°C (41°F) and 40°C (104°F).

However, studies undertaken by the US Department of Energy compared the hygrothermal performance of real walls, assessed through both monitoring and visual inspection, with the predicted performance calculated according to the standard. The results showed that ASHRAE 160-2009 criteria were providing false positives and were highly conservative compared to actual performance. In addition, other studies have demonstrated that the values of the data contained in the table 'Residential design moisture generation rates' presented by the standard were too high, which resulted in an overestimation of the moisture load on the building components (Glass & TenWolde, 2009).

Therefore a new version of the standard was released to address these problems. To improve the reliability of the code, the moisture performance evaluation adopted alternative mould-growth criteria (the VTT Mould Growth Index) to indicate risks based on a sliding scale of mould growth instead of the binary pass/fail criterion defined by the ASHRAE 160-2009.

7.2.2 General comments

The intent of this standard is to supersede the usual prescriptive approach to hygrothermal assessment in favour of a performance-based procedure, enabled by new computer-based moisture analysis tools (refer to Chapter 6: Hygrothermal modelling).

As described in the previous chapters, excessive indoor humidity and moisture can impact the durability of materials and occupants' health. However, the correlations between cause and effect are extremely complex, due to the high number of parameters involved, such as mould species, temperature and substrate, to name a few. ASHRAE 160-2016 seeks to mitigate this complexity and provides a code that is easy to apply by defining a clear pathway for the assessment and referring to updated research findings. In terms of accuracy, the model's reliability is negatively affected by its inability to account for the contribution of airflow to moisture transfer within the envelope. Indeed, even a small airflow is capable of transferring more moisture than the amount that can be transported by vapour diffusion. However, because this phenomenon is difficult to model, calculate and predict, the standard identifies it as optional. The weather file represents another shortcoming of the standard: the MDRY is currently not available for most locations. This can lead to the use of incorrect weather data that can significantly affect the reliability of the results. To mitigate this risk, ASHRAE undertook a research project that developed guidelines and examples for the generation of MDRY (Salonvaara, 2011).

Despite the current limitations, ASHRAE 160-2016 is an international reference for standards and regulations in regard to moisture control and is the leading international hygrothermal policy framework.

7.3 Building Code of Australia

The Building Code of Australia (BCA) (Australian Building Code Board, 2019a) first acknowledged and sought to address the problem of condensation in its latest revision, released in 2019.

This important update mainly involves the following documents:

- BCA Volume 1 (Focus on Section F—Health and amenity);
- BCA Volume 2 (Focus on Part 3.8 Health and amenity);
- NCC 2019 BCA Guide (Focus on Part F6 Condensation management);
- Handbook Condensation in buildings (Version 2019);
- Condensation in Buildings—Tasmanian Designers' Guide—Version 2.

7.3.1 Introduction

The NCC acknowledges that the issues related to condensation are complex and it is hard to find a simple solution suitable for all cases. It explicitly states that 'the intent of these requirements is to assist in the mitigation of condensation within a building. The installation of a condensation management system may not prevent condensation from occurring' (Australian Building Code Board, 2019a).

Although the Building Code defines 10 different classes of buildings, based on their use, the condensation risk requirements are provided for only three of these classes, all of which relate to residential buildings:

- in a sole-occupancy unit of a Class 2 building (Volume 1);
- Class 4 part of a building (Volume 1);
- Class 1 Building (Volume 2).

The BCA 2019 guidelines consider a residential building to be more susceptible to moisture accumulation due to its intended function, the unpredictable nature of its use, and the rapid uptake of the latest architectural trends in the Australian residential market. The increasing requirements for insulation and airtightness and the shift to air-conditioned homes are changing the indoor environment and contributing to the increased risk of condensation and mould growth (Australian Building Code Board, 2019b).

Typically, Australian residential buildings rely heavily on natural ventilation; they are rarely designed to integrate mechanical ventilation systems. High performing and sealed envelopes, as required by the current increasingly stringent regulations, are rarely accompanied by updated ventilation strategies and rely solely on operation by the building occupants. However, it is highly unlikely that occupants will open windows to ventilate the indoor area if the AC is running, which increases the risk of condensation due to the increment in indoor humidity and the different conditions (in terms of RH and temperature) between indoors and outdoors. The "Handbook: condensation in buildings', which is a non-prescriptive government document, identifies other causes of the increased risk of condensation in the Australian built environment:

- Increased water vapour resistance in claddings, linings and finishes, such as plastic and metal claddings, rendering of brickwork and vinyl wallpapers, which can reduce drying potential through walls.
- Wider use of engineered timber products with lower moisture storage capacity and less mould resistance than the solid timber sections they replace.
- Greater use of lightweight construction, also reducing the capacity of the building envelope fabric to detain moisture for later drying.
- Reduced emphasis on effective ventilation of roofs and other interstitial spaces to dilute water vapour concentrations.
- Reduced infiltration and air leakage through the building envelope as a result of improved sealing, reducing the opportunities for drying and increasing indoor water vapour levels.
- Lower ventilation levels in dwellings no longer occupied during the day, reducing opportunities for drying and increasing indoor water vapour levels.
- Rapidly growing use of air conditioners for summertime cooling in temperate and cooler climates, creating out-of-season condensation risks and reversing the expected direction of water vapour flows during the course of the year.

7.3.2 Scope and verification methods

The Australian code discusses condensation risks in the section on 'Health and Amenity' (Australian Building Code Board, 2019a), identifying the main goal as the minimisation of condensation-related issues for the building occupants' health. The Guide to NCC 2019 Volume 1 clearly acknowledges the connections between condensation, mould growth and poor indoor environmental quality. However, it also recognises that the impacts are not limited to health and well-being but can also affect structural integrity and cause building degradation in general, due to interstitial condensation.

The goal of the hygrothermal assessment is clarified in section FV6, which indicates that the primary aim is to ensure that moisture will not accumulate inside the primary water control layer or on its interior surface.

Consistent with the approach typically adopted in the BCA, Part F6 "Condensation Management section' allows for two different pathways for the designer/professional to design and certify the building:

1. the 'deemed-to-satisfy' (DTS) method, which reports prescriptive provisions specific to the membrane typology and position within the building envelope;
2. the performance solution method, which allows compliance with the performance requirements to be demonstrated according to the criteria established by the code in sections A2.2(3) and A2.4(3).

7.3.3 The deemed-to-satisfy solution

The DTS approach, as the name suggests, aims to achieve a compliant design solution through a series of provisions that focus on minimising the risk of condensation and mould growth through a multilevel approach. First, it aims to minimise the internal RH via specific requirements for kitchen and bathroom, sanitary compartment and laundry (F6.3). Second, it aims to improve the dissipation of vapour through the use of ventilated roof (F6.4) or drained cavity (F6.2), and pliable membranes (water barrier) located on the external side of the insulation.

For warmer climate zones (climate zones 1–5), where heat and vapour migration are from outside to inside, the maximum risk of condensation is on the external side of the membrane. Here an impermeable external water membrane addresses the waterproofing performance while reducing the risk of interstitial condensation. This solution also increases the drying potential as it ensures a quick and easy drying out of the condensation, taking advantage of the proximity of the condensate to the exterior high temperatures.

Colder climates (zones 6, 7 and 8) are characterised by a prevailing heating mode, which increases the risk of interstitial condensation on the warmer side of the water barrier (between the membrane and the insulation) as a result of the migration of warm and humid internal air towards the external side of the building element. In this case, the code prescribes the use of permeable pliable building membrane, defined according to the relevant code (AS/NZS 4200.1, 2017) to allow the vapour to dissipate and reduce the risk of condensation.

7.3.4 Performance solution

The DTS provisions are often criticised as being insufficiently flexible. Accordingly, the code provides an alternative pathway to certify compliancy: the performance solution method. As established by the Australian code, a compliant performance solution must consider variables and objectives defined in FV6, it must be undertaken by an expert and it must show clear evidence of suitability, verification method and comparison with the DTS solution. The code does not provide more specific details or cite references to the method or tool to be used to undertake the assessment.

Only the 'Handbook: Condensation in buildings', produced by the Australian Building Codes Board, provides guidance on this topic. It contains an overview of the main international codes and a framework to be used for the performance solution method. Its preface notes that the Handbook provides 'non-mandatory advice and guidance' (Australian Building Code Board, 2019c). Section 6 of the Handbook describes different methods of condensation assessment compliant with the 'Performance solution criteria' as described by the BCA (Volumes 1 and 2). These methods are divided into steady state and transient:

1. Simplified steady-state method: BS EN ISO 13788:2012: Hygrothermal performance of building components and building elements—internal surface temperature to avoid critical surface humidity and interstitial condensation.

2. Transient conditions method: BS EN 15026:2007: Hygrothermal performance of building components and building elements—assessment of moisture transfer by numerical simulation. The Handbook identifies WUFI as one of the most advanced software used in this field.

Like BS 5250:2011, the Handbook provides guidelines on the boundary conditions and assumptions to be used in the assessment. Section 4.2 gives some indications in regard to the generation of the weather file, suggesting that the most severe climate in the previous 10/20/50 years be used. The Handbook also gives the minimum parameters that should be included in the climate file:

- dry-bulb temperature
- vapour pressure, or any other humidity parameter that can be used to calculate vapour pressure
- global and diffuse solar radiation
- sky temperature
- wind speed and direction
- total atmospheric pressure
- precipitation (rain, snow, drizzle)

In regard to internal climate generation, it suggests the use of ANSI/ASHRAE Standard 160.

7.3.5 General comments

Compared to the previous codes analysed in this chapter, it is clear that the scope of the provisions in the Australian NCC is limited, as it refers only to new residential buildings and ignores different building types and remedial work. Indeed, the NCC 2019 code addresses the problem of condensation and moisture management with an overly simplified approach that is limited to the definition of the degree of permeability of the external 'pliable membrane' depending on the climate zone. In this regard, further limitations are presented. As explained in Chapter 4, Health and mould growth, all studies show that the growth of mould is dependent not only on temperature but also on water activity and, therefore RH. Considering that the eight NCC climate zones are only 'developed with an emphasis on defining the desirable thermal characteristics of building envelopes', it is possible that they are not 'reliable indicators of condensation risk in buildings' (Australian Building Code Board, 2019b). To address these problems, an alternative definition of zones based on outdoor overnight condensation potential is being explored and might become a more reliable tool for design and assessment in the future.

Another issue relates to the management of condensation related to thermal bridges. The NCC condensation section (Volumes 1 and 2) makes no reference to the relationship between condensation and thermal bridges. Only the Handbook provides an adequate explanation, although it mainly refers to other international standards. The BS 5250:2011 and BS EN ISO 13788:2012 address this problem in a more comprehensive way, providing references that explain how to perform the required assessments.

One of the reasons for the present gap between the Australian code and the other two codes is related to climate. Australia is characterised by a wide range of climates, which poses unique challenges. In fact, research and expert discussion on moisture management and mould growth in temperate or hot and humid climates are rare. Focused research is needed to address this knowledge gap to improve this aspect of the Australian code.

7.4 Concluding remarks

The topics of condensation, moisture buffering, moisture management and mould growth are all interconnected and constitute a relatively new topic of interest in the field of building physics, applied research and, ultimately, building performance policy. The increasing need for high-performance buildings and comfortable internal environments characterised by a high level of air quality is highlighting the present and future risks related to airtight envelopes, mould growth and durability issues. A growing body of research is providing policymakers with more and more reliable information, contributing to constant improvement of the codes. However, gaps in the existing knowledge can disrupt this process and increase the distance between state-of-the-art practice and building codes. This appears to be the case in Australia. Although moving in the right direction, the unparalleled diversity of the climate and the lack of locally specific research have resulted in a building code that relies on a few incomplete prescriptive indications that fail to offer reliable guidance on how to avoid the risks of condensation and mould growth.

All building standards are increasingly recognising the complexity and the gravity of the problem of mould growth and acknowledging the negative impacts in terms of durability, economics, health and safety. Although it appears that all codes are slowly moving towards a more scientific performance/simulation approach to the design of quality internal spaces, all the national standards considered in this review still include many references to prescriptive approaches, benchmark solutions or, in some cases, to prepackaged construction details as primary resources to reduce the risk of condensation/mould growth. As the analysis has shown, software simulation is always considered as a secondary option (in case the prescriptive approach is not applicable) or as a 'non-mandatory' measure. As discussed in relation to ASHRAE 160-2016, where the code was recently updated to better reflect the real-case scenario, one possible reason for this is the continued lack of confidence in the results provided by the simulations, as a result of past and recent examples of false positives and overly conservative results. Another factor could be the complexity of the problem of mould growth, which translates into complex advanced simulation methods and software that require specialised and experienced input. Most of the time, the design and construction of small residential buildings is financially driven and involves a small team. Accordingly, there is limited capacity to use these advanced tools, and prescriptive solutions are more likely to be effective.

Tables 7.1 and 7.2 show, respectively, a summary of the building codes and standards reviewed in this chapter, and a comparison between the assessment methods defined in BS EN ISO 13788:2012, BS EN 15026:2007 and ASHARE 160-2016.

Table 7.1 A comparison between the condensation assessment methods described in the different building codes analysed.

Code	Typology				Storage and transport phenomena considered					Scope		
	1D	2D	Steady state	Non-steady state	Convection/ air movement	Heat and moisture buffering	Water movement for capillarity	Solar radiation	Moisture penetration due to driving rain	Definition of boundary conditions	Calculation formulas/ method	Definition of satisfactory performance
BS EN ISO 13788:2012	X		X*							X	X	
BS EN 15026:2007	X			X		X		X	X	X	X	
ANSI / ASHRAE 160-2016	X**	X**		X		X	X	X	X	X	***	X

*This standard also provides a steady-state methodology to assess the drying potential calculated over one year.
**This standard refers to software that can provide 1D or 2D results.
***This standard refers to software with inbuilt calculation processes.

Table 7.2 A comparison between the building code and standards highlighting the differences and limitations.

Reference country	Code/ standard	Nature of the document	Building typologies				Approach			Recipients		
			All building typologies	Only residential	New built	Renovation	Prescriptive indications	High-level recommendations	Calculation/ simulation methodology	Designers	Construction companies	Building user and owners
United Kingdom	BS 5250:2011	National Building code	X		X	X	X	X	*	X	X	X
	BS EN ISO 13788:2012	International standard	X		X	X			X	X		
	BS EN 15026:2007	International standard	X		X	X			X	X		
United States	ANSI/ASHRAE 160-2016	International standard	X		X	X			X	X		
Australia and Tansmania	BCA 2019	National Building code		X	X		X			X		
	Handbook Condensation in Buildings 2019	Good practice	X		X	X		X	*	X	X	X

*Reference to other codes.

Australia seems to have the most limited building code, in which only one type of building (new residential building) is considered. The Handbook could bridge the gap in the regulatory framework, but it remains a non-mandatory standard for designers. The set of codes included in and referenced by the UK Building Code represents the most thorough attempt to provide guidance to designers, builders and occupants in regard to moisture-related risk management. In this regard, it must be noted that renovations and new constructions may suffer from different issues and peculiarities and that it is of utmost importance to provide reliable and robust guidelines specific for each case.

It is clear that, with the advent of transient 2D tools, the steady-state Glaser approach has become obsolete as an assessment method, since its usability and reliability are negatively affected by a long list of limitations and assumptions. Over the past decade, the accuracy of the results of hygrothermal simulations has been greatly improved by state-of-the-art transient hygrothermal software, together with the latest standardised framework (ASHRAE 160-2016) and postprocessing models (VTT). The method used by ANSI ASHRAE 160-2016 clearly helps to reduce the limitations of the assessment and introduces more accurate and standardised assumptions that generate more robust results. The shortcomings of the postprocessing VTT model in terms of materials and climate, and the lack of a model or software that takes account of air convection, continue to be limiting factors for these new methodologies and present opportunities for future research.

References

AS/NZS 4200.1 (2017). Pliable building membranes and underlays materials.

ASHRAE 160. (2016). Criteria for moisture-control design analysis in buildings.

ASTM E96. (2016). Standard test methods for water vapor transmission of materials.

Australian Building Code Board. (2019a). Building code of Australia.

Australian Building Code Board. (2019b). Condensation in building handbook.

Australian Building Code Board. (2019c). Handbook: Condensation in buildings.

BS 5250. (2011). Code of practice for control of condensation in buildings.

BS EN 15026. (2007). Hygrothermal performance of building components and building elements. Assessment of moisture transfer by numerical simulation.

BS EN ISO 13788. (2012). Hygrothermal performance of building components and building elements. Internal surface temperature to avoid critical surface humidity and interstitial condensation.

Building Regulations and Approved Documents. (n.d.). Retrieved November 25, 2020, from https://www.gov.uk/government/collections/approved-documents.

Clark, N., Ammann, H. M., Brunekreef, B., Eggleston, P., Fisk, W., Fullilove, R., Guernsey, J., Nevalainen, A., & Essen, S. G. V. (2004). Damp indoor spaces and health.

EN ISO 15927-4. (2005). Hygrothermal performance of buildings—Calculation and presentation of climatic data—Part 4: Hourly data for assessing the annual energy use for heating and cooling.

Glass, S. V., & TenWolde, A. (2009). Review of moisture balance models for residential indoor humidity. In *Proceedings of the 12th Canadian conference on building science and technology*, 1, 231–245.

IBC. (2018). International building code.

Lacy, R. E. (1965). Driving-rain maps and the onslaught of rain on buildings: Building Research Station.

May, N., & Sanders, C. (2016). Moisture in buildings. https://sdfoundation.org.uk/downloads/BSI-White-Paper-Moisture-In-Buildings.PDF.

Salonvaara, M. (2011). RP-1325 - Environmental weather loads for hygrothermal analysis and design of buildings.

TenWolde, A. (2011a). A review of ASHRAE Standard 160—Criteria for moisture control design analysis in buildings. In *Condensation in Exterior Building Wall Systems*.

TenWolde, A. (2011b). A review of ASHRAE Standard 160—Criteria for moisture control design analysis in buildings. In *Condensation in Exterior Building Wall Systems*.

TenWolde, A., & Walker, I. S. (2001). Interior moisture design loads for residences. In ASHRAE (Ed.), *Performance of exterior envelopes of whole buildings VIII [electronic resource]: Integration of building envelopes*.

Glossary

Absolute humidity Mass of vapour per unit volume in the sample of air under any given condition (Hens, 2016).

Absorption coefficient (of water) (A_w) A coefficient that quantifies the amount of water which, when in contact with a porous material, enters that material due to absorption. It is defined by the following equation:

$$m_s = A_w \times \sqrt{t} \tag{1}$$

where m_s is the mass of sorbed moisture per unit of contact area (kg/m^2) and t is time in second (Hagentoft, 2001; ISO EN 9346:2007, 2007)

Adsorption The adhesion of atoms or molecules from gas or liquid to a surface.

Air change rate, also referred to as ventilation rate (n) A measure of the airflow rate through an enclosed volume.

$$n = \frac{R_s}{V} \tag{2}$$

with R_s airflow rate (m^3/h) and V enclosed ventilated volume (m^3).

Air content (w_a) Mass of air per unit volume of material (kg/m^3).

Airflow resistance (S_a) Ratio between the air pressure difference across the bounding surfaces under 1-D steady-state conditions and the density of the airflow rate

$$S_a = \frac{p_1 - p_2}{r_a} \tag{3}$$

where r_a is the density of airflow rate either [m^3/(m^2 s)] or [kg/(m^2 s)], and p_1, p_2 the air pressure on each side of the layers (Pa) (Hagentoft, 2001; ISO EN 9346:2007, 2007).

Air leakage rate Volume of air movement per unit time across the building envelope. Air leakage is uncontrolled and accidental and is most often the result of poor design or construction methods. Air leakage should not be confused with ventilation which, although measured with the same units, is controlled and deliberate. Air leakage can have a variety of sources, such as joints, cracks and porous surfaces. This air movement through the building envelope is generally caused by a difference of pressure between the internal and external space, due to either mechanical pressurisation and depressurisation or natural wind pressures, or air temperature differentials.

Air ratio (X_a) Mass of air per unit mass of dry material (%) or (kg/kg) (Hens, 2016).

Air saturation degree (S_{air}) Ratio between the air content and the maximum admissible; can also be described as the percentage of pores filled with air in relation to all those accessible (%) (Hens, 2016).

Analytical model A mathematical model that has a closed-form solution to the governing equations. Compared to numerical models, analytical models and the mathematical function with which they are associated can provide information about the system's behaviour without the need for graphing or generating a table of values (Delgado, Barreira, Ramos, & Freitas, 2012).

Building component Part of a building that can be identified as an independent subsystem or subassembly due to its specific function (e.g. wall, structure) or its separate manufacturing process.

Building envelope The external skin of a building, consisting of all the components that provide a boundary or barrier that separates the building's interior volume from the outside environment or from different interior environments. An interior partition that separates two dissimilar environments, such as a cold storage facility adjacent to an occupied office, can be treated as a building envelope element for modelling purposes (Hens, 2017).

Building physics The science that applies the principles of physics to the built environment.

Bulk density Mass divided by the volume occupied by a material.

Capillary suction The difference in pressure resulting from capillary forces that induce a liquid to enter a porous medium.

Conduction The process by which heat is transferred through the collision of vibrating adjacent atoms. In the presence of one or more solids with different temperatures, heat is transferred by conduction from the higher to the lower temperature area.

Convection The process by which something is transferred through the movement of groups of molecules at different temperatures.

Critical surface humidity Relative humidity at the surface that leads to deterioration of the surface, including mould growth (%) (ISO EN 13788:2012, 2012).

Degradation factor (or agent) External factors that affect the performance of building materials and components. Examples of degradation factors are weathering, biological stress and incompatibility between materials and use (wear and tear) (Masters & Brandt, 1989).

Degradation mechanism The sequence of chemical, biological, mechanical or physical mechanisms that result from exposure to one or more degradation factors and which has a negative impact on one or more properties of a building material or component (Masters & Brandt, 1989).

Density (δ) Mass per unit volume of a material (kg/m^3). In a homogeneous material, the density is a constant. In heterogeneous materials, the density varies between different regions but in construction is usually averaged and assumed as a constant.

Design life (of a building material or component) The period of time after installation during which a product is expected to meet all its minimum functional requirements when routinely maintained with minor or major repairs.

Deterioration The process of becoming impaired in quality, performance, aesthetics or value (Masters & Brandt, 1989).

Dew point temperature The temperature at which vapour pressure reaches saturation value. Condensation occurs at any temperature below the dew point.

Dry-bulb temperature Air temperature measured in a thermometer shielded from radiation (ISO EN 15927:2005, 2005).

Durability The capability of a building, assembly, component, product, material or construction to maintain serviceability over a specified time. Durability depends on the inherent properties of the material or component and its interaction with the environment.

Dynamic viscosity (η) The ratio of viscous shear stress and the velocity gradient in the normal flow (sPa).

Emissivity (ε) The ratio between the energy emitted by a real surface and the total existence of a black body at the same temperature (Hagentoft, 2001; ISO 9288:1989, 1989).

Enthalpy (H) The sum of the system's internal energy and the product of pressure and volume of a thermodynamic system.

Enthalpy flow The process by which heat is transferred by the movement of a gas or fluid permeating a medium. This process is linked to convection (CIB - W040, 2012; ISO EN 9346:2007, 2007).

Finite difference method (FDM) A numerical method that uses a discretisation technique to linearise governing partial differential equations (PDE) by approximating derivatives with finite differences (Delgado et al., 2012).

Finite control volume method (FDM) This method discretises the space occupied by the variables into controlled volumes and conservatively evaluates the flux at the boundary between volumes. This method is usually used in fluid dynamics or applied to physical conservation laws (Delgado et al., 2012).

Finite element method (FEM) A type of approximate, numerical model that divides the domain of the solution into a discrete number of nonoverlapping elements represented by a finite number of mesh or grid points (nodes), where the functions are approximated to local functions (such as polynomials), resulting in a system of equations that describes the whole problem (Delgado et al., 2012).

Glaser method A simplified method of calculating the amount of condensate deposited and/or evaporated over a 12-month period. This method is limited to 1 dimension and is used to quantify the maximum amount of accumulated moisture due to interstitial condensation caused by vapour flow through the building element.

Growth model A mathematical function/formula that aims to describe the growth (positive or negative) of a specific entity in dependence on various boundary conditions.

HAM Acronym for heat, air and moisture. It is usually used to identify models that allow assessment of the transport processes of heat, air and moisture in a building and building envelope.

Heat flow (φ) Movement of heat per unit of time (J/s), (W) (Hens, 2016).

Heat flux (q) Movement of heat per unit of time across a unit of surface normal to the flow (W/m^2).

Heat transfer Energy transfer by a combination of conduction, convection, radiation and enthalpy flow (Hagentoft, 2001; ISO EN 13788:2012, 2012).

High performing building envelope Envelope designed to minimise the thermal energy between internal and external spaces.

Humidity ratio (x) Mass of water vapour within a unit mass of dry air. At saturation, it is represented as x_{sat} (kg/kg) (Hens, 2016).

Hydrolysis Chemical process of decomposition in which a molecule of water ruptures one or more chemical bonds.

Hygrothermal model A mathematical model that describes a building envelope system or subsystem, defining approximate thermal and moisture transport mechanisms and initial-boundary system performance under applied conditions. Several hygrothermal models can be used to describe the monodimensional or multidimensional hygrothermal performance of a building envelope or a whole building (Delgado et al., 2012).

Hygroscopic range The range of relative humidity in a material between 0% and 98% (Hens, 2016).

Hyphae A long chain of cells that make up a fungus (Torvinen et al., 2006).

Hysteresis Phenomenon in which the state of a physical system depends not only on its present inputs but also on past inputs.

Interior/Internal volume The deliberately conditioned space within a building; it generally excludes attics, basements and attached structures such as garages, unless these are connected to the heating and air conditioning system.

Isopleth A curve that expresses the correlation between temperature, humidity and exposure time to determine mould growth. An isopleth system is a diagram that expresses favourable growth conditions in relation to temperature and relative humidity (Vereecken & Roels, 2012).

Maintenance The group of activities performed during the design life of a building that aims to maintain the performance of a component/assembly (minor or routine maintenance) or to reinstate the performance of a component/assembly to fulfil its intended function (major maintenance).

Mass transfer Transport of mass by various mechanisms. The term also applies to the movement of moisture or air (Hagentoft, 2001; ISO EN 9346:2007, 2007).

Moisture Presence of small amounts of water. The term moisture transfer refers to all movements of liquid and vapour across the building envelope and the whole building, including diffusion, convection, wind pressure, capillary suction and gravity.

Moisture capacity Change in the mass of moisture in unit mass [specific moisture capacity (kg/kg Pa)] or volume [volumetric moisture capacity (kg/m^3 Pa)] that follows from unit change in capillary suction. Moisture capacity is also defined as the slope of the sorption isotherm and is specific for different materials.

Moisture content (*w*) Mass of moisture per unit volume of material (kg/m^3) (Hens, 2016)

Moisture driving potential The main factors responsible for determining the formation of moisture under specific conditions.

Moisture effusivity (b_m) It is a theoretical parameter that can be used to indicate the rate with which moisture is absorbed by a material when subjected to variations in surface humidity. Moisture effusivity can be calculated as:

$$b_{\mathrm{m}} = \sqrt{\frac{\delta_{\mathrm{V}} \cdot p_0 \cdot \frac{\partial V}{\partial RH}}{p_{\mathrm{sat}}}} \tag{4}$$

where δ_V [kg/(msPa)] is the vapour permeability, p_0 (kg/m^3) is the density of the sample, $\partial v/\partial RH$ is the relation between the moisture content and the relative humidity and p_{sat} (Pa) is the saturation pressure (Rode et al., 2006).

Moisture ratio (χ) Mass of moisture per unit mass of dry material (%), (kg/kg) (Hens, 2016).

Moisture saturation degree (*S*) The ratio between the moisture content and the maximum admissible (%), also described as the percentage of pores filled with moisture in relation to all those accessible (Hens, 2016).

Mould The microfungi that colonise buildings are colloquially called mould (Leung & Lee, 2016).

Mould index The index that identifies the intensity of mould growth on the surface of a material. In the VTT model, the mould index comprises levels of mould growth risk (from 1 to 6), where the higher the index, the higher is the risk (Oreszczyn & Pretlove, 2000).

Moisture reference year (MRY) A year of hourly weather data that have been selected for use in hygrothermal analysis. The file collects all relevant information that is useful to characterise the external climate of a specific geographical location. An MRY includes hourly data on:

- dry-bulb temperature (°C)
- relative humidity (%)
- wind velocity m/s and direction, counted clockwise from north over east
- rain (mm)
- ambient air pressure (hPa)
- direct or global solar radiation on the horizontal plane (W/m^2)
- diffuse solar radiation on the horizontal plane (W/m^2)
- cloud index.

An MRY differs from a standard methodological file because it contains more stringent conditions, which allows it to take into consideration variation in climate and worst-case scenarios. The literature presents different statistical methods to generate this dataset, based on the statistical analysis of monitored meteorological data (Salonvaara, Zhang, & Karagiozis, 2011).

Mycelium The microfungi commonly called mould produce spores that are naturally found in the air and on various materials and which, under favourable conditions, can grow and germinate to form a fungal mass, called mycelium (Leung & Lee, 2016).

Numerical model A model that uses numerical methods to solve the governing equations of the applicable problem. When applied to building physics and physics in general, the model establishes the physical representativeness of the solution, and the numerical method defines the accuracy to which the model can be approximated.

Onset of mould growth The moment at which it becomes possible to see some germination activity under the microscope (Isaksson, Thelandersson, Ekstrand-Tobin, & Johansson, 2010).

Partial vapour pressure (P_{vap}) The partial pressure expected by vapour in the volume occupied by a mixture of gases (air), at a determined temperature (T). Considering atmospheric air as the sum of dry air and water vapour, it can be stated that the atmospheric pressure exerted by the air is the sum of the partial pressures of dry air and water vapour: (Hens, 2016; Luikov, 1964)

$$P_{tot} = p_{vap} + p_{air} \tag{5}$$

Porosity Also referred to as total porosity, this is the volume of pores per unit volume of material expressed in (m^3/m^3). The open porosity Ψ_0 is the volume of open pores per unit volume of material (m^3/m^3). In general, the open porosity is lower than the total porosity $\Psi_0 < \Psi$.

Psychrometric chart A psychrometric chart graphically approximates the properties of moist air, expressing its thermodynamic parameters at a constant atmospheric pressure.

Radiation The process by which heat is transferred by the emission and absorption of electromagnetic waves. Since every surface with a temperature higher than 0 K emits electromagnetic energy, if surfaces have different temperatures, heat exchange occurs.

Relative humidity (RH or ϕ) Ratio between the vapour pressure and the saturated vapour pressure at a certain temperature (%) (ISO EN 13788:2012, 2012)

$$\mathrm{RH} = \frac{p}{p_{\mathrm{sat}}} \tag{6}$$

Service life Also called design service life of a building material or component, this is the period of time after installation during which a product is expected to meet all its minimum functional requirements when routinely maintained but without major repair becoming necessary (Masters & Brandt, 1989).

Sorption Process that includes the effect of both absorption and adsorption.

Sorption curve This represents the moisture content in a porous material and the relative humidity of the ambient air at equilibrium obtained at precisely controlled temperature and relative humidity. It identifies two different regions: the hygroscopic range, usually up to RH 98%, and the capillary area. The sorption curve can be empirically determined by following the ASTM Standard C1498 (Hagentoft, 2001; ISO EN 13788:2012, 2012; ISO EN 9346:2007, 2007).

Specific heat capacity The heat required to raise the temperature (usually one degree) of the unit mass of a given substance by a given amount.

Steady state Condition of a system where all relevant parameters do not change (or are assumed not to change) with time.

Suction curve A curve that expresses the relation between the moisture content in a porous material and suction in pore water. This relation includes hygroscopic region (sorption curves) and above hygroscopic region (Hagentoft, 2001; ISO EN 9346:2007, 2007).

Temperature factor (f_{Rsi}) Difference between the temperature of a surface indoors and the external air temperature, divided by the difference between the air temperature indoors and the air temperature outdoors (ISO EN 13788:2012, 2012)

$$f_{\mathrm{Rsi}} = \frac{\theta_{\mathrm{si}} - \theta_{\mathrm{e}}}{\theta_{\mathrm{i}} - \theta_{\mathrm{e}}} \tag{7}$$

Design temperature factor ($f_{\mathrm{Rsi,\ min}}$) At the internal surface. It is defined as the minimum acceptable temperature factor at the internal surface to avoid mould growth and other durability issues on building materials (ISO EN 13788:2012, 2012)

$$f_{\mathrm{Rsi,min}} = \frac{\theta_{\mathrm{si,min}} - \theta_{\mathrm{e}}}{\theta_{\mathrm{i}} - \theta_{\mathrm{e}}} \tag{8}$$

where $t_{\mathrm{si,min}}$ is the minimum acceptable internal surface temperature before mould growth can start, t_{i} is the internal operative temperature and t_e is the external air temperature.

Thermal bridge A localised area in the building envelope where there is a higher heat flow compared to a standard section of the envelope. Thermal bridges are usually caused by a localised reduction in the amount of insulation due to the presence of a noninsulated component such as a structural member or a mechanical connection (structural thermal bridge). Alternatively, thermal bridges can occur where the insulation space is reduced due to the geometry of the building, such as corners and edges (geometric thermal bridges). Thermal bridges can be local (1D), linear (2D) or diffuse (3D) and can be quantified through the calculation of the localised U-value (3D) (W/m^2 K), Ψ-value (2D) (W/m K), or X-value (1D) (W/K). Thermal bridging is not only detrimental to the energy performance of the envelope but can also contribute to lowering the internal surface temperature, increasing the chances of condensation with consequent mould growth (Hagentoft, 2001; Hens, 2016).

Thermal conductivity (λ) A measure of the intrinsic property of a material to transfer heat. It is defined as the density of heat flow rate per one unit of the thermal gradient in the direction of the flow. Thermal conductivity depends on the material's composition, density, temperature and moisture content [W/(m K)] (ISO 7345:1987, 1987).

Thermal conductance A measure of the capacity of a layer/assembly to transfer heat. It is defined as the thermal conductivity divided by the thickness of the layer [W/(m^2 K)] (ISO 7345:1987, 1987).

Thermal resistance (*R*) This is defined as the reciprocal of the thermal conductance [(m^2 K)/W] (ISO 7345:1987, 1987).

Time of wetness (TOW) The period during which a surface is covered by adsorptive and/or liquid films of electrolyte that can cause atmospheric corrosion. The definition of TOW is also dependent on the definition of "wet" and how to measure it. Numerous practical and theoretical definitions have been developed over the past 60 years. In this book, TOW is defined as the ratio between the wet period (RH above 80%) and the total time period considered (Johansson, Bok, & Ekstrand-Tobin, 2013).

Typical meteorological year (TMY) Datasets consisting of 1 year of hourly weather data from long-term data records (at a minimum, 10 years). They provide a relatively concise dataset from which system performance estimates are developed. Since the TMY data are generated by determining "typical" meteorological months, they do not provide information on extreme events and do not necessarily represent actual conditions at any given time, but they are useful to compare different design solutions, thus limiting the amount of required computations (Renné, 2016).

Total pressure Dalton's law defines the total pressure of a mixture of gases as the sum of the partial pressure of each gas. Considering atmospheric air as the sum of dry air and water vapour, it can be stated that the atmospheric pressure exerted by the air is the sum of the partial pressures of dry air and water vapour (Hens, 2016):

$$p_{tot} = p_{vap} + p_{air} \tag{9}$$

Transient hygrothermal simulation software Software for the dynamic simulation of heat and moisture transfer in a building component. Transient hygrothermal software usually runs hourly based simulations, expressing the outputs as hourly values.

Vapour barrier/retarder A material or layer specifically intended to restrict the transmission of vapour through a building component.

Vapour concentration (*ν* or ρ_v) The mass of vapour per unit of air volume (kg/m^3). According to the ideal gas law, the vapour concentration is a function of the partial vapour and temperature:

$$\rho_V = \frac{m_V}{V} = \frac{p_{sat}}{R \times T} \tag{10}$$

At saturation, the notations ν_{sat}, $\rho_{v,sat}$ are used (Hens, 2016).

Vapour permeability (δ_v) A measure of the intrinsic characteristic of a material to allow vapour to pass through it. It is defined as the quantity of water per unit area that can be diffused in a unit of time, calculated for a specific gradient of relative humidity. It can be expressed as permeance per unit of thickness. The vapour permeability of different materials is quantified through empirical tests, such as the traditional cup-test [kg/(m s Pa)] (Hagentoft, 2001; ISO EN 9346:2007, 2007).

Vapour permeance The density of vapour transmission rate through a layer of a given thickness, over the course of one unit of time, at a specified vapour pressure difference across two parallel surfaces under steady-state conditions. Similar to thermal conductivity and thermal conductance, the main difference between vapour permeability and vapour permeance is that the latter depends on the thickness of the material while the former does not (Perm), [kg/(m^2 s Pa)] (ISO EN 13788:2012, 2012).

Vapour resistance factor (μ) A measure of the intrinsic tendency of a material to prevent water vapour passing through. It is also defined as the reciprocal of vapour permeance (Hagentoft, 2001; ISO EN 9346:2007, 2007).

Volumetric air ratio (Ψ_a) The volume of air per unit volume of material (%), (m^3/m^3) (Hens, 2016).

Volumetric moisture ratio (Ψ) The volume of moisture per unit volume of material (%), (m^3/m^3) (Hens, 2016).

Water activity (a_w) The vapour pressure on the substrate divided by that of pure water at the same temperature:

$$a_w = \frac{p}{p^*} \tag{11}$$

where p is the vapour pressure on the substrate and p^* is the vapour pressure of pure water at the same temperature (Nielsen, Holm, Uttrup, & Nielsen, 2004).

Water vapour diffusion-equivalent air layer thickness (s_d) The thickness of a motionless air layer that has the same water vapour resistance as the material layer in question (ISO EN 13788:2012, 2012):

$$s_d = \mu \times s \tag{12}$$

Weather barrier A material or system that stops the ingress of air and water under gravity and pressure gradient.

Weathering factor Environmental factors such as radiation, temperature, rain and other forms of water, freezing and thawing (Masters & Brandt, 1989).

Wind-driven rain (WDR) The rain transported not only by gravity but also by the movement of air due to the presence of wind. It can be expressed as:

$$R_{WDR} = R_h \frac{U}{V_t} \tag{13}$$

with R_h intensity of rainfall falling through a horizontal plane (m^3/m^2), U wind speed in m/s and v_t raindrop terminal velocity of fall in m/s.

References

CIB - W040. (2012). In V. P. De Freitas, & E. Barreira (Eds.), *Heat, air and moisture transfer terminology: parameters and concepts* (369).

Delgado, J. M., Barreira, E., Ramos, N. M., & Freitas, V. P. (2012). *Hygrothermal numerical simulation tools applied to building physics*. Springer.

Hagentoft, C. E. (2001). *Introduction to building physics*. Studentlitteratur.

Hens, H. S., & Hugo, S. L. (2016). Applied building physics. Ernst & Sohn.

Hens, H. S. (2017). *Building physics-heat, air and moisture: fundamentals and engineering methods with examples and exercises*. Wiley.

Isaksson, T., Thelandersson, S., Ekstrand-Tobin, A., & Johansson, P. (2010). Critical conditions for onset of mould growth under varying climate conditions. *Building and Environment*, *45*(7), 1712–1721. Available from https://doi.org/10.1016/j.buildenv.2010.01.023.

ISO 7345:1987. (1987). Thermal insulation - Physical quantities and definitions.

ISO 9288:1989. (1989). Thermal insulation - Heat transfer by radiation - Physical quantities and definitions.

ISO EN 15927:2005. (2005). Hygrothermal performance of buildings - Calculation and presentation of climatic data.

ISO EN 9346:2007. (2007). Hygrothermal performance of buildings and building materials - Physical quantities for mass transfer – Vocabulary.

ISO EN 13788:2012. (2012). Hygrothermal performance of building components and building elements - Internal surface temperature to avoid critical surface humidity and interstitial condensation - Calculation methods.

Johansson, P., Bok, G., & Ekstrand-Tobin, A. (2013). The effect of cyclic moisture and temperature on mould growth onwood compared to steady state conditions. *Building and Environment*, *65*, 178–184. Available from https://doi.org/10.1016/j.buildenv.2013.04.004.

Leung, M. H., & Lee, P. K. (2016). The roles of the outdoors and occupants in contributing to a potential pan-microbiome of the built environment: A review. *Microbiome*, *4*(1), 1–15.

Luikov, A. V. (1964). Heat and mass transfer in capillary-porous bodies. *Advances in Heat Transfer*, *1*(C), 123–184. Available from https://doi.org/10.1016/S0065-2717(08)70098-4.

Masters, L. W., & Brandt, E. (1989). Systematic methodology for service life prediction of building materials and components. *Materials and Structures*, 22(5), 385–392. Available from https://doi.org/10.1007/BF02472509.

Nielsen, K. F., Holm, G., Uttrup, L. P., & Nielsen, P. A. (2004). Mould growth on building materials under low water activities. Influence of humidity and temperature on fungal growth and secondary metabolism. *International Biodeterioration and Biodegradation*, *54*(4), 325–336. Available from https://doi.org/10.1016/j.ibiod.2004.05.002.

Oreszczyn, T., & Pretlove, S. (2000). *Mould index. Cutting the Cost of Cold*. E&FN Spon, (pp. 122-133).

Renné, D. S. (2016). *Resource assessment and site selection for solar heating and cooling systems. Advances in solar heating and cooling*. Woodhead Publishing.

Rode, C., Peuhkuri, R., Lone, H., Time, B., Gustavsen, A., Ojanen, T., … Arfvidsson, J. (2006). *Moisture buffering of building materials. NT Technical Report 592*. Nordtest.

Salonvaara, M., Zhang, J., & Karagiozis, K. (2011). *Enivronmental Weather Loads for Hygrothermal Analysis and Design of Buildings. ASHRAE RP-1325. Simulation studies and dataanalysis*.

Torvinen, E., Meklin, T., Torkko, P., Suomalainen, S., Reiman, M., Katila, M. L., ... Nevalainen, A. (2006). Mycobacteria and fungi in moisture-damaged building materials. *Applied and Environmental Microbiology*, *72*(10), 6822–6824. Available from https://doi.org/10.1128/AEM.00588-06.

Vereecken, E., & Roels, S. (2012). Review of mould prediction models and their influence on mould risk evaluation. *Building and Environment*, *51*, 296–310. Available from https://doi.org/10.1016/j.buildenv.2011.11.003.

Index

Note: Page numbers followed by '*f*' and '*t*' refer to figures and tables, respectively.

P

R

S

T

U

V

W